Breakthrough! : how
greatest discoveries
10643345

Breakthrough!

How the 10 Greatest Discoveries in Medicine Saved Millions and Changed Our View of the World

Jon Queijo

Vice President, Publisher: Tim Moore
Associate Publisher and Director of Marketing: Amy Neidlinger
Editorial Assistant: Pamela Boland
Acquisitions Editor: Kirk Jensen
Development Editor: Russ Hall
Operations Manager: Gina Kanouse
Senior Marketing Manager: Julie Phifer
Publicity Manager: Laura Czaja
Assistant Marketing Manager: Megan Colvin
Cover Designer: Stauber Design Studio
Managing Editor: Kristy Hart
Project Editor: Anne Goebel
Copy Editor: Language Logistics
Proofreader: Kathy Ruiz
Indexer: Erika Millen
Graphics: Laura Robbins
Senior Compositor: Gloria Schurick
Manufacturing Buyer: Dan Uhrig

© 2010 by Pearson Education, Inc.
Publishing as FT Press Science
Upper Saddle River, New Jersey 07458

FT Press Science offers excellent discounts on this book when ordered in quantity for bulk purchases or special sales. For more information, please contact U.S. Corporate and Government Sales, 1-800-382-3419, corpsales@pearsontechgroup.com. For sales outside the U.S., please contact International Sales at international@pearson.com.

Company and product names mentioned herein are the trademarks or registered trademarks of their respective owners.

All rights reserved. No part of this book may be reproduced, in any form or by any means, without permission in writing from the publisher.

Printed in the United States of America

First Printing March 2010

ISBN-10: 0-13-713748-6
ISBN-13: 978-0-13-713748-0

Pearson Education LTD.
Pearson Education Australia PTY, Limited.
Pearson Education Singapore, Pte. Ltd.
Pearson Education North Asia, Ltd.
Pearson Education Canada, Ltd.
Pearson Educación de Mexico, S.A. de C.V.
Pearson Education—Japan
Pearson Education Malaysia, Pte. Ltd.

Library of Congress Cataloging-in-Publication Data

Queijo, Jon, 1955-

Breakthrough! : how the 10 greatest discoveries in medicine saved millions and changed our view of the world / Jon Queijo.

p. cm.

Includes bibliographical references.

ISBN 978-0-13-713748-0 (hardback : alk. paper) 1. Medicine—Research—History. 2. Discoveries in science. I. Title.

R852.Q45 2010

610.72—dc22

2009051043

With love and gratitude to my parents,
Anthony A. Queijo and June Dudley Queijo.

And a very special thanks to Rooshey Hasnain,
who helped me every step of the way with her many
ideas and her boundless encouragement, support,
and caring.

"The most exciting phrase to hear in science, the one that heralds the most discoveries, is not 'Eureka!' but 'That's funny.'"

—*Isaac Asimov, American writer and biochemist*

"That's funny..."

—*Alexander Fleming, on finding a moldy growth that led to his discovery of penicillin*

Contents at a Glance

Contents

Acknowledgments

Thanks to all of the folks at Pearson/FT Press Science for their hard work and encouragement in bringing this book to fruition, especially Tim Moore, Russ Hall, Gina Kanouse, Anne Goebel, Julie Phifer, Megan Colvin, Kirk Jensen, Amanda Moran, Chrissy White, Laura Robbins, and Pam Boland. And finally, thanks to Jim Markham, for introducing me to the world of book writing.

About the Author

Jon Queijo has been writing about science and medicine for more than 25 years in positions that include senior medical writer in the pharmaceutical industry, staff writer for the *New England Journal of Medicine* consumer publication *Weekly Briefings*, and staff writer for *Bostonia* magazine. His freelance articles have appeared in various publications, including *Brain Work* (The Dana Foundation), *Psychology and Personal Growth*, *Environment*, and *Science Digest*. He has a Bachelor of Science degree in technical journalism from the University of Florida and a Master of Science degree in science communication from Boston University.

Introduction

It's tempting to start right off with an apology for the word "breakthrough," a word that—depending on your point of view—can be as tedious as an overhyped headline or seductive as a brightly wrapped gift. Either way, it's hard to resist wondering, *What* breakthrough? A cure for cancer, an easy way to lose weight, the secret to living forever? But this isn't *that* kind of book, and apologies seem unnecessary when you're talking about the ten greatest breakthroughs in *all* of medicine. Sadly, none involve easy weight loss or living forever. However, all are arguably *more* important because they meet three essential criteria: 1) They saved, improved, or reduced suffering in millions of lives; 2) They changed the practice of medicine; and 3) They transformed our understanding of the world. That last item is too often overlooked. All medical "breakthroughs" profoundly impact health and how physicians work; far rarer are those that open our eyes to a fundamentally new way of seeing the world, giving new meaning to not only such questions as, Why do we get sick, and how do we die? but also, How are we put together and what connects us to the rest of nature?

Each of these ten breakthroughs came at a time in history when they struck humanity like a thunderbolt—a jolt of awakening followed by a palpable rise in human consciousness. *What?* Illness is caused by natural forces and *not* evil spirits or angry gods? Inhaling certain gases can take away pain and *not* kill the patient? A machine can take pictures of the *inside* of your body? We often take it for granted today, but at one time, millions of people couldn't believe what they were hearing. They *refused* to believe it. Until they finally did. And then the world would never be the same.

Critics often have a field day with top ten lists. Motives are immediately suspect, every selection second-guessed, many "better" alternatives offered. But objectivity tends to crumble when one tries to measure how much any one discovery has impacted suffering, illness, and death. Nevertheless, one valid criticism is that top ten lists are overly simplistic. In our celebrity-obsessed times, the spotlight's glare on a handful of superstars too often blinds us to the many individuals who helped pave the way. Yet what often makes a great discovery most fascinating is understanding the many smaller steps that often made the final "leap" possible. This book celebrates these steps and shows how—milestone by milestone—they led to ten final breakthrough discoveries.

Don't embark on these journeys expecting tales of calculating genius and easy success. In fact, the greatest breakthroughs in medicine represent a wildly unpredictable collage of human stories and emotion. Even if you're not surprised to learn how many discoveries relied on one individual's dogged persistence despite failure and repeated rejection, you may be shocked to learn how many discoveries resulted from sheer dumb luck, if not divine intervention. The number of "coincidences" that led to Alexander Fleming's discovery of penicillin might tempt some atheists to reconsider their assumptions. Also surprising is how many individuals had no idea their work would one day lead to a major breakthrough. One example is Swiss physician Friedrich Miescher, who discovered DNA in 1869—more than 70 years before scientists would figure out its role in heredity.

But though ignorance in the pursuit of truth is forgivable, it's harder to sympathize with those throughout history who ridiculed a discovery because fear and rigid thinking prevented them from letting go of outdated beliefs and tradition. The examples are many: from the rejection of the pioneering work in germ theory by John Snow and Jacob Semmelweis in the early 1800s to the dismissal of Gregor Mendel's laws of genetics in the 1860s when—despite ten years of hard work—one eminent scientist snorted that Mendel's work had "really just begun." No doubt, many of the greatest discoveries in medicine were made by courageous individuals who dared shake the foundations of a long-held, and usually wrong, view of the world. And no surprise that, once the discovery was finally accepted and solid footing regained, the world found itself in a very different place.

* * *

And yet the nagging question remains: why *these* ten breakthroughs and *these* rankings? If you have other ideas and some free time on your hands, you could try creating your own list by, for example, typing "medical breakthrough" into Google. However, you might want to set aside the morning to narrow down your choices from the *2.1 million* hits that came up in one search conducted in 2009. Fortunately, my task was simplified by a 2006 poll conducted by the *British Medical Journal* (*BMJ*) in which readers were asked to submit their nominations for the greatest medical breakthrough since 1840 (the year the *BMJ* was first published). After receiving more than 11,000 responses, votes were narrowed down to a final 15.

Entries that did *not* make the *BMJ*'s final 15 ranged from the disingenuous (plastic, the iron bedstead, the tampon, Viagra, and the welfare state) to the sincere (blood tests, the defibrillator, clot-busting drugs, insulin, nurses, and caring for the terminally ill). However, the final 15 had a curious feel of being both idiosyncratic *and* (mostly) right: 1) sanitation (clean water and sewage disposal); 2) antibiotics; 3) anesthesia; 4) vaccines; 5) discovery of DNA structure; 6) germ theory; 7) oral contraceptives; 8) evidence-based medicine; 9) medical imaging (such as X-rays); 10) computers; 11) oral rehydration therapy (replacement of fluids lost through vomiting and diarrhea); 12) risks of smoking; 13) immunology; 14) chlorpromazine (first antipsychotic drug); and 15) tissue culture.

The *BMJ*'s top 15 is fine but hardly the final word. Another list published in 1999 by the Centers for Disease Control and Prevention's publication *Morbidity and Mortality Weekly Report (MMWR)* offered an interesting twist with its "Ten Great Public Health Achievements" in the United States from 1900 to 1999. The *MMWR* did not rank its selections but shared some similar items with the *BMJ* list (vaccines and "control of infectious diseases"), while offering some other valid entries, including improvements in motor vehicle and work place safety, safer and healthier foods, declines from heart disease and stroke, and recognition of tobacco use as a hazard.

While the *BMJ* and CDC lists influenced my selection of the top ten breakthroughs, both had limitations, such as excluding medical breakthroughs before 1840 (a decision Hippocrates and a few others might take issue with). In addition, it seemed more interesting and relevant to morph the *BMJ*'s "discovery of chlorpromazine" into the more inclusive "Medicines for the Mind." As described in Chapter 9, this breakthrough covers one of the most remarkable ten-year periods in the history of medicine: From 1948 through the 1950s, scientists discovered drugs for the four most important mental disorders to afflict the human race: schizoprenia, manic-depression, depression, and anxiety.

Another question likely to arise when considering the "top ten" of anything is: What's *that* doing there? For example, many people associate "medical breakthroughs" with various technological marvels (MRI imaging, lasers, artificial body parts), surgical feats (organ transplants, tumor removal, angioplasty), or miracle drugs (aspirin, chemotherapy, cholesterol lowering agents). Yet, while one can point to numerous examples in each of these categories, none rank among the top ten when the previously mentioned criteria are considered. In fact, it's interesting to note that two of the *BMJ*'s top 15 rankings are decidedly *low*-tech: sanitation (#1) and oral rehydration therapy (#11). Yet both are clearly high-yield in terms of lives saved. It's estimated, for example, that over the last 25 years, oral rehydration therapy has saved the lives of some 50 *million* children in developing countries.

Along the same lines, others may argue *against* some of the breakthroughs included here, such as the rediscovery of alternative medicine. I'm thinking of one former editor of the *New England Journal of Medicine* who declined to review this book in part because "There is no such thing as 'alternative medicine'—only medical methods that work and those that don't." I understand the point but respectfully disagree. There are many ways of addressing the pros and cons of alternative medicine—some of which I hope are reasonably covered in Chapter 10, "A Return to Tradition." However, when all factors are considered from a larger perspective—a very large canvas that covers pretty much all of human history—I stand by its inclusion.

The first and easiest explanation for including alternative medicine is to point to the partnerships now forming between alternative medicine and scientific medicine and the recent birth of a new philosophy of healing that draws on the best practices of *both* traditions. This new and rapidly developing field, called "integrative medicine," is now supported by many practitioners in both alternative and scientific medicine. A second point is that even lacking the methodology of scientific medicine, alternative medicine has had a remarkably positive effect on the health and spirits of millions of people who don't necessarily subscribe (or prescribe) to a purely Western scientific model of health and disease. But a third and perhaps more important point is that taking a closed-minded view toward nonconventional approaches to medicine has an uncomfortable feel to it, a queasy reminder of lessons learned throughout the history of medicine, as taught to us by those who rejected William Harvey's theory of circulation, the value of René Laennec's stethoscope, Edward Jenner's vaccine for smallpox, the theory that germs can cause disease, Mendel's laws of genetics, the value of ether in surgery, the idea that penicillin can stop bacterial infections, that…

Well, you get the point.

* * *

Perhaps the best thing about the top ten breakthroughs in medicine is the stories they reveal about people from all walks of life—physicians, scientists, patients, and ordinary folk. The stories cover a wide range of emotion, from the disbelief and awe of witnessing a deep secret of nature suddenly revealed, to the relief and joy of discovering a new tool that saves patients from pain and certain death. But always, they are the stories of how the human spirit pushes the boundaries of knowledge in new and surprising ways, such as:

- Hippocrates, who invented clinical medicine with his painfully detailed observations of patients like the boy from Meliboea, who died a slow and agonizing death "as a result of drunkenness and much sexual indulgence."
- Physician Ignaz Semmelweis, who returned from vacation to learn that he'd lost his close friend to a disease that was supposed to only affect women and who was subsequently struck by an insight that would save countless lives.
- The youths in the early 1800s who inhaled experimental gases and indulged in "frolics" and "jags," unaware that their experiences would help pave the way to the discovery of anesthesia.
- Farmer Benjamin Jesty who, 20 years before Edward Jenner "discovered" vaccines, led his family into a cow pasture and vaccinated them against small pox based on a hunch and an old wives' tale.

It's often moving to read the stories of people who couldn't imagine how their efforts and suffering would one day impact millions of lives and change our view of the world. Ironically, we're just as much in the dark today every time we learn of a new discovery, whether in the news or among the two million hits that come up in a Google search. No one can say which will stand as a true breakthrough two years, much less two centuries, from now. Quite possibly, it will be tomorrow's discovery of a new cancer cure, easy weight loss, or infinite longevity. In the meantime, here are ten that we *know* have stood the test of time. Without them, we might not have the luxury of such speculation—or perhaps of having been born at all.

The World's First Physician: HIPPOCRATES AND THE DISCOVERY OF MEDICINE 1

The Greek island of Kos, located in the crystal clear waters of the Aegean Sea and bordered by 70 miles of golden beaches, might be one of the best spots on the planet to fall ill—or to simply remain well.

Part of a 12-island archipelago, Kos is 200 miles southeast of Athens and just a few miles off the southwest coast of Turkey. Long, narrow, and verdant with lush foliage, the island is flat except for two low mountains along its southern coast. But it is in the town of Kos, an ancient village on the northeastern coast of the island, where the magic and medicine of this island begin.

It is tempting to speculate that the legendary history of Kos arises from its nurturing rich soil and abundant groundwater: Visitors entering the village are greeted by a lush landscape of tall palms, cypresses, pine trees, jasmines and, for an added splash of color, the bright reds, pinks, and oranges of the hibiscus. But if you want to locate the true pulse of Kos and its 2,500-year-old legacy, you must continue your journey…

First, face west and walk two and a half miles out of the village. There, amidst more lush landscape, you will approach a sloping site. Hiking up this slope, you pass an extensive complex of ancient ruins that rise around you in a series of terraces. Put aside your curiosity and continue to climb. Before long, you will arrive at a pinnacle. Gazing out from this high point, you stop flat in your tracks: The world has split apart.

Spread out before you is a breathtaking view of the Aegean seacoast. Inhaling the fresh coastal air, you feel the stirrings of the true spirit of this small island, the mystery of where two worlds meet. One, the "inner" world, is simply you: the tightly wrapped sac of blood and bone, emotion and mind, which is your body. The other "outer" world is merely every other thing in the physical universe surrounding you.

If you ponder for a moment the possibility *that two such worlds not only exist, but* co-exist *in a place we may not yet fully understand, then congratulations. You have finally begun to arrive, physically and metaphysically, on the island of Kos. For this is the place where, in the view of the world's first "rational physician," all life, death, health, and disease—and hence the practice of medicine and healing itself—begins.*

* * *

This ancient site is known as the *Asklepieion*, the generic Greek word for any "healing temple." But the Asklepieion of Kos is a temple like no other. Although today a crumbling ruin of broken walls, roofless chambers, and lonely columns supporting only air, in its heyday this was a bustling center of healing. Here patients in all stages of sickness and injury sought the best treatment they could find.

If you were suffering from disease or injury and arrived here in the fifth century BC, over the course of days and weeks you would have progressively ascended each of the four terraces that scale these grounds, each level dedicated to a different stage of diagnosis, counseling, and healing. Apart from simple relaxation, your treatment might have included bathing in large pools, being massaged with perfumes, oils, and ointments, following a regimen of mental and physical exercise, receiving diet counseling, herbs and other oral drugs, and commiseration with the ancient spirits.

Oh, and one more thing. If you happened to check in sometime between 490 and 377 BC, you might have received one more benefit: a visit from the world's "first" physician, a man not only credited with inventing the practice of medicine, but whose insights have remained influential for well over two thousand years.

* * *

Most of us have a distinct yet vague impression of who Hippocrates was. The phrase "Father of Medicine" often (and accurately) jumps to mind. And of course, there is the Hippocratic Oath, which we know has something to do with doctors behaving nicely. On the other hand, it should be noted that Hippocrates bears no connection to the similar-sounding "hypocrisy." Though also Greek in origin, hypocrisy is from *hypokrisis*, which means "playing a part" or, as commonly used today, someone who is a phony.

Which Hippocrates certainly was not.

Who was Hippocrates, and how did earn the mantle of "Father of Medicine" as well as credit for the "Invention of Medicine"?

Perhaps one measure of an individual's greatness is when the question is not so much how their "breakthrough" compares to another, but rather, *which* of their many breakthroughs one should select to make the comparison. For Hippocrates, the list is substantial and includes being the first physician to

- Recognize that diseases have natural causes, rather than their arising from supernatural or evil forces
- Invent "clinical medicine" and the "doctor-patient relationship"

- Create an oath of conduct that has remained influential for 2,500 years
- Elevate the practice of medicine to an honored profession, rather than a conventional trade like plumbing or roof repair
- Achieve many other medical breakthroughs, including recognizing that thoughts and emotions arise in the brain rather than the heart

And yet...

* * *

Some time around 440 BC a young physician eager for knowledge crossed the narrow body of water separating his island home from what we know today as southwestern Turkey. Reaching land, he made his way 50 miles north to an area known as Ionia. Entering the city of Miletus, he met with the well-known philosopher Anaxagoras. Famed for introducing philosophy to the Athenians, Anaxagoras was also the first person to recognize that the moon's brightness is due to reflected light from the sun. The ensuing conversation must have been interesting. On the one hand, Hippocrates was a reputed descendent of Asklepius, the god of healing and son of Apollo. On the other hand, Anaxagoras was likely unimpressed by religious tradition: In 450 BC he had been imprisoned for insisting that the sun was not a god. While this outrageous claim may have raised the hackles of any other healer from Kos, more likely it set a twinkle in the eye of young Hippocrates. And an invitation to sit down for a chat...

* * *

And yet among the many "firsts" commonly attributed to Hippocrates, one breakthrough at the core of his teachings is often forgotten or overlooked today. Perhaps this lapse is due to its paradoxical nature, the fact that it both opposes yet resonates with the way medicine is often practiced today. What was this additional breakthrough? Before answering, we need to learn more about this man and his place in history.

The making of the man: 19 generations of healers and 3 first-rate legends

In today's high-tech world of CAT, MRI, PET, SPECT, and other cryptic visions, of the increasing specialization and molecularization of medicine, of all manner of pharmacopeia from the fitful to the fatal, we put a certain amount of trust in the rituals of modern medicine. We are comforted by hospital rooms where patients are anchored to their sanitized beds by the wires and tubes of modern technology. If for some reason then, you were to succumb to an illness and wake up in the fifth century BC in a dim, oil-lamp-lit chamber to the sound of a priest moaning incantations over your hurting body, chances are you would be overcome by a distinct lack of confidence, if not terror.

Hippocrates may well have felt the same.

Yet, born on Kos in 460 BC, this was the world in which he was raised. Like many doctors today, Hippocrates came from a line of physicians who had been practicing "medicine" for generations. For starters, he was trained in medicine by his father, Heracleides, his grandfather, and other famous teachers of the time. But this is being too modest. In fact, his family also claimed that the tradition of medicine had been in their lineage for no less than 19 generations, dating back to Asklepieios, the demi-god of healing. Deities aside, Hippocrates' early view of medicine was probably influenced by a long, long ancestry of religious healers and priests.

If you're thinking that claiming to be the nineteenth-generation descendent of the god of healing on your medical school application might strain the limits of credulity—or, conversely, that it might be just the edge you need for acceptance—several caveats are in order. First, surprisingly few undisputed details are known about the life of Hippocrates. Although a large body of writings attributed to Hippocrates have survived—some 60 works collectively known as the *Corpus Hippocraticum*, or simply Hippocratic Corpus—there is considerable debate as to which are genuine works of Hippocrates versus the embellishments of the many admirers who expanded on his school of thought decades and even centuries after his death. Nevertheless, by comparing and analyzing the documents, historians have patched together a reasonably credible account of Hippocrates and his accomplishments.

* * *

To be honest, three of the most colorful stories about Hippocrates are probably rooted as much in legend as they are truth. But even if only partly true, they provide insight into the man Hippocrates may well have been, a man whose reputation was sufficiently formidable to spread beyond his own small island to the distant lands of his own enemies.

The first and perhaps best known story is set in 430 BC during the Peloponnesian War. Shortly after being destroyed by the Spartans, a plague broke out in the city of Athens. Hippocrates and his followers traveled to Athens to help. Observing that the only people not affected by the plaque were iron smiths, Hippocrates made an astute deduction: Their resistance must somehow be related to the dry, hot atmosphere in which they worked. He promptly wrote up his prescription. The citizens of Athens were to light fires in every home to dry the atmosphere, to burn corpses, and to boil all water before consumption. The plague retreated, and Athens was saved.

The second story is often cited to highlight Hippocrates' remarkable diagnostic skills, which ranged from the physical to the psychiatric. Shortly after the Athenian plague, King Perdiccas of Macedonia, aware of Hippocrates' growing reputation, requested the physician's help when no other doctor could diagnose his vexing symptoms. Hippocrates agreed and traveled to Macedonia to see the king. During the examination, Perdiccas blushed whenever a beautiful girl named Phila—who was his father's concubine—was nearby. Hippocrates took note. Upon further inquiry, he learned that Perdiccas had grown up with Phila and dreamed of one day marrying her. This dream was shattered when his father took the girl as his concubine. However, the recent death of his father reawakened Perdiccas' conflicted feelings of love for Phila, causing him to fall ill. After subsequent counseling by Hippocrates, the king was cured.

The third story, a testament to Hippocrates' loyalty, took place when Greece was at war with Persia. By this time, Hippocrates' reputation was so great that Artaxerxes, the enemy king of Persia, requested that Hippocrates travel to Persia to save its citizens from a plague. Despite the king's offer of gifts and wealth "equal to his own," Hippocrates politely declined. Although sympathetic, it was against his scruples to assist the enemy of his country. The king gracefully responded with a vow to destroy the island of Kos—a threat that was put to rest, figuratively and literally, when the king suffered a stroke and died.

Legends aside, our investigation into the invention of medicine may be better served by looking at the achievements of Hippocrates as documented in the more scholarly writings of the Corpus. While historians continue to debate the authenticity of even these documents, said caveats having been acknowledged, we can venture into the territory where Hippocrates' "Invention of Medicine" can be attributed to six major milestones.

And yet...

* * *

While there is no record of the conversation between Hippocrates and Anaxagoras in that ancient city of Miletus, it's not hard to imagine that the young physician was beginning to question the medical tradition of his own family, with its lineage of demi-gods, superstitions, and priest healers. It wasn't that Hippocrates completely rejected their theocratic approach; he simply felt that in medicine and health, other truths were to be found. Thus, the reputation of Anaxagoras and his philosophy, which had reached even the small island of Kos, brought Hippocrates here to question and learn. Settling down in the shade of a tree outside the city, Hippocrates proffered a simple invitation. "You know of my background and tradition, Anaxagoras. Now tell me of yours..."

Milestone #1

Getting real: diseases have natural causes

> "[Epilepsy] appears to me to be nowise more divine nor more sacred than other diseases... Men regard its cause as divine from ignorance and wonder."
>
> —*Hippocrates,* On the Sacred Disease, *420-350 BC*

Until the time of Hippocrates, the most commonly accepted explanation for the cause of essentially all illness was refreshingly simple: Punishment. Having been found guilty of some misbehavior or moral failure, the gods or evil spirits exacted their justice through sickness. Your redemption, or "treatment" as we call it today, might include a visit to a nearby temple of Asklepieios, where the local priests attempted to cure your malady with incantations, prayers, or sacrifice.

At some point early in his career, Hippocrates changed the rules. Distancing himself from the Asklepian priests and their theocratic approach to healing, Hippocrates insisted that diseases were caused by natural forces and not the gods. No statement better summarizes his view than the frequently quoted passage from one of the books attributed to Hippocrates, *On the Sacred Disease*. The title of this book—the first to be written about epilepsy—references the belief at the time that seizures were caused by the "sacred" hand of a displeased god.

Hippocrates begged to differ:

> "It appears to me to be nowise more divine nor more sacred than other diseases, but has a natural cause like other affections. Men regard its nature and cause as divine from ignorance and wonder, because it is not at all like other diseases. And this notion of its divinity is kept up by their inability to comprehend it… Those who first referred this disease to the gods appear to me to have been just such persons as the conjurers and charlatans… Such persons, then, using divinity as a pretext and screen of their own inability to afford any assistance, have given out that the disease is sacred…"

In this and similar writings, we hear in the voice of Hippocrates not only the adamancy of his view that disease arises from natural causes, but the exasperation, if not contempt, he holds for the "charlatans" who would claim otherwise. Thus with such statements, and nothing more than his own mortal powers, Hippocrates wrestled disease from the supernatural and placed it squarely in the world of the rational and natural.

Milestone #2

It's the patient, stupid: the creation of clinical medicine

> "His symptoms were shivering, nausea, insomnia, and lack of thirst… He was delirious, but calm, well-behaved and silent."
>
> —*Hippocrates,* Epidemics 3, *420-350 BC*

The term "clinical medicine" embodies much of what we now like to think any good doctor practices. It includes everything from taking a detailed patient history, to performing a careful physical examination and recording of symptoms, to diagnosis, treatment, and an honest assessment of the patient's response to that treatment. Prior to Hippocrates, the practitioners of medicine were not especially concerned with such details. Rather than focusing on the pains and woes of individual patients, early Greek physicians tended to take a one-size-fits-all approach, in which patients were subjected to ritualistic, predetermined, and highly non-individualized treatments. In changing that approach, Hippocrates founded the art and science of clinical medicine.

How does one invent "clinical" medicine? Some say that Hippocrates developed his clinical insights through exposure to a long and curious tradition in the Asklepieion of Kos. For many years, patients recovering from illness would inscribe in the temple an account of the help they had received so that it might be useful for other patients. According to this story, Hippocrates took on the task of writing out these inscriptions and, armed with this body of knowledge, established the practice of clinical medicine.

More likely, the clinical skills developed by Hippocrates and his followers were earned through hard work over the course of many years and many interactions with many patients. One vivid and typical example of these skills, recorded in the book *Epidemics 3*, involves a youth in Meliboea who apparently was no icon of Greek virtue. According to Hippocrates, the youth "had been feverish for a long time as a result of drunkenness and much sexual indulgence… His symptoms were shivering, nausea, insomnia, and lack of thirst." Although not for the squeamish, the subsequent description of the youth's demise demonstrates a skill of clinical observation that could stand as a model for any medical student today:

> "*First day*: There passed from his bowels a large quantity of solid stools with much fluid. During the following days he passed a large quantity of watery, greenish excrement. His urine was thin, sparse, and of bad color. His respiration was at long intervals and deep after a time. There was a flabby tension of the upper part of the abdomen extending laterally to both sides. Cardiac palpitation was continuous throughout... *Tenth day*: He was delirious, but calm, well-behaved and silent. Skin dry and taut; stools either copious and thin or bilious and greasy. *Fourteenth day*: All symptoms exacerbated. Delirious with much rambling speech. *Twentieth day*: Out of his mind; much tossing about. No urine passed; small amounts of fluid retained. *Twenty-fourth day*: Died."

Through such clinical observation—with its focus on individual patients and their symptoms—Hippocrates raised medicine from the dusky gloom of demons and rituals into the bright light of keen observation and thought. And it made perfect sense in the world view that Hippocrates had begun to shape: if diseases had natural cases, why not look more closely at symptoms for clues as to what those causes might be? What's more, this new focus on individual patients paved the way for another component that we now regard as essential to good medicine: the "doctor-patient relationship."

Milestone #3

A code of ethics that stands the test of time

> "I will use treatments for the benefit of the sick to the best of my ability and judgment; I will abstain from doing harm or wronging any man by it."
>
> —*Hippocrates,* Oath, *420-350 BC*

Among all the known writings from Antiquity, the *Hippocratic Oath* is considered by some to be second in authority only to the *Bible*. Adopted as a code of behavior by physicians throughout history, the *Oath* continues to influence many physicians today and is still frequently cited in scholarly journals and the popular press as *the* code of ethics for the proper practice of medicine.

Contained in a single page of text, the *Oath* begins with the physician swearing, "by Apollo the physician, and Asklepius…and all the gods and goddesses as my witnesses" to uphold the *Oath* and its contract. In subsequent statements, the physician is bound to uphold a variety of ethical and behavioral standards, including:

- Holding my teacher "equally dear to my parents" and being willing to "impart a knowledge of the art to my own sons"
- "Not giving a lethal drug to anyone if I am asked"
- "Avoiding any voluntary act of impropriety or corruption, including the seduction of women or men, whether they are free men or slaves"
- Keeping secret "whatever I see or hear in the lives of my patients, whether in connection with my professional practice or not."

Although some biographies suggest that Hippocrates required his apprentices to swear by an oath before he would accept them as students, the origin of the *Oath* as we know it today is unclear and may have been rewritten a number of times over the ages to suit the needs of different cultures. In any case, the *Oath* was hardly the last word by Hippocrates on ethics and the proper practice of medicine. For example, in the book *Epidemics*, he offers one of his best-known maxims—one that most patients today would be happy to remind their doctors of while being trundled into the operating room:

> "Regarding diseases, make a practice of two things—to help or, at least, do no harm."

Milestone #4 Acting the part: professionalizing the practice of medicine

> "He must be clean in person, well-dressed, and anointed with fragrant perfumes that do not in any way cause suspicion…"
>
> —*Hippocrates,* Physician, *420-350 BC*

Living in the twenty-first century AD, it is difficult to imagine how healers in the fifth century BC conducted their daily business. However, it seems reasonable to assume that between the priests and their incantations and various peripatetic healers with their non-FDA approved ointments, the practice of medicine was fairly loose by today's standards. In various books and writings, Hippocrates changed this, too. Raising the practice of medicine from a common trade to a profession with rigorous standards, he provided advice in virtually every arena of medicine.

For example, recognizing that not everyone is cut out for medical training, Hippocrates cautions in one book:

> "Whoever is going to acquire truly an understanding of medicine must possess the following advantages: natural ability, instructions, a suitable place for study, tuition from childhood, industry, and time. First of all, natural ability is required, for, if nature is in opposition, all is in vain."

In another text, he describes a range of physical and personality traits physicians need to possess to successfully practice medicine:

> "The authority of a physician requires that he is of healthy complexion and plump as nature intended... Next, he must be clean in person, well-dressed, and anointed with fragrant perfumes that do not in any way cause suspicion."

In another text, however, Hippocrates cautions against the perils of vanity:

> "You must also shun luxurious headgear with a view to procuring patients, and elaborate perfume too."

What's more, the physician must be mindful of demeanor and the appropriate boundaries of laughter. "In appearance he must have a thoughtful but not harsh countenance; for harshness seems to suggest stubbornness and misanthropy. But, on the other hand, the man of uncontrolled laughter and excessive cheerfulness is considered vulgar. Such a disposition must especially be avoided."

And what patient today would not be reassured by Hippocrates' formula for bedside manner?

> "When you enter a sick man's room...know what you must do before going in... On entering, be mindful of your manner of sitting, and your reserve, your decorum, authoritative demeanor, brevity of speech, your composure, your response to objections, and your self-possession in the face of troublesome occurrences."

As for the occasional troublemaker, Hippocrates advises,

> "It is necessary also to keep an eye on the patient's faults. They often lie about taking the things prescribed [and] die through not taking disagreeable potions."

Despite his stern advice, Hippocrates' underlying goodwill is unmistakable:

> "Give what encouragement is required cheerfully and calmly, diverting his attention from his own circumstances. On one occasion rebuke him harshly and strictly, on another console him with solicitude and attention."

And finally, when it comes to the sensitive issue of billing, Hippocrates reveals a spirit both sympathetic…

> "One ought not be concerned about fixing a fee. For I consider an anxiety of this sort harmful to a troubled patient. It is better to reproach patients whom you have saved than to extort money from those who are in a critical condition."

and charitable…

> "Have regard to your patient's means or wealth. On occasion, give your services free, recalling the memory of an earlier debt of gratitude…"

Milestone #5 The enigmatic Corpus: 60 books and a wealth of medical firsts

> "Men ought to know that the source of our pleasures, merriment, laughter, and amusements as well as our grief, pains, anxiety and tears is none other than the brain."
>
> —*Hippocrates,* On the Sacred Disease, *420-350 BC*

Much of what we know of the medicine of Hippocrates comes from the Hippocratic Corpus, a collection of about 60 manuscripts that covers virtually every aspect of health, from the inner (mind and body), to the outer (environment), to where the two worlds meet (diet and breathing).

Although the Corpus as we know it today dates back to 1526, a mere 500 years ago, accounting for its whereabouts in the preceding 2,000 years is a bit more problematic. Some historians believe that the manuscripts were initially assembled in the Great Library at Alexandria around 280 BC, possibly after they were recovered from the remains of the medical school library at Kos.

What *else* do we know about these manuscripts? On the perplexing side, their hodge-podge of mixed content, writing styles, chronology, and contradictory viewpoints suggests that they were written by multiple authors who lived before and after Hippocrates. On the other hand, though none of the writings can be definitively linked to Hippocrates, most were probably written around 420 BC to 350 BC, corresponding to his lifetime. Most intriguing, despite a pervasive lack of inner unity, the manuscripts share one crucial theme: a belief in rationality and a scorn for magic and superstition.

To get an idea of why historians are vexed in their attempts to make any generalizations about the Corpus, one needs only to consider the curious diversity of their titles, which include: *Nature of Man*; *Breaths*; *Nutriment*; *Aphorisms*; *Dentition*; *Airs, Waters, and Places*; *Affections*; *Joints*; *On Diseases*, *Decorum*; *Head Wounds*; *The Nature of the Child*; *Diseases of Women*, and so on. And the content ranges wildly in form and content, from a series of easily memorized sentences (*Dentition*), to insightful medical observations (*On the Sacred Disease*), to simple lists of ailments (*On Diseases*).

Nevertheless, from these texts we can gather that Hippocrates and his followers had a remarkably accurate understanding of anatomy—perhaps derived from their observations of war wounds and animal dissections—given that at the time, human dissections were deemed unacceptable, if not forbidden. True, at times the descriptions tended to lean a bit heavily on analogy and metaphor—for example, the eye was compared with a lantern and the stomach to an oven. But in other cases their anatomical and clinical observations were so accurate that they have earned the admiration of physicians and surgeons throughout history, up to and including the twenty-first century.

Some of the most fascinating observations from the Corpus come from facts that we take for granted today but were quantum leaps of insight at the time. One of the best examples is Hippocrates' descriptive assertion in *On the Sacred Disease* that thought and emotion arise from the brain and not the heart, as others believed at the time:

> "Men ought to know that the source of our pleasures, merriment, laughter, and amusements, as well as our grief, pains, anxiety, and tears, is none other than the brain. It is by this organ that we think, see, hear, and distinguish between the ugly and the beautiful... By this same organ, too, we become mad or delirious, and are assailed by fears and panics, by insomnia and sleepwalking..."

Among the anatomical and clinical descriptions that continue to impress physicians today are those describing head injuries and joint deformities. For example, some claim that Hippocrates' treatise *On Injuries of the Head* helped set the stage for modern-day neurosurgery. The treatise begins with an impressively detailed discussion of the anatomy of the skull, including cranial structure, thickness, and shape, and differences in texture and softness between the skulls of adults and children. Hippocrates then describes six specific types of cranial trauma, including fissured fractures (caused when a weapon breaks the bone), depressed fractures, and wounds above cranial sutures. Other details reveal his clinical experience in treating head injuries, such as his description of certain cranial fractures that are "so fine that they cannot be discovered...during the period in which it would be of use to the patient."

Similar details of medical acumen are seen in the manuscript *On Joints*, in which Hippocrates describes techniques for managing spinal diseases, including correction of curvatures of the spine and spinal injuries. Particularly interesting is the Hippocratic table, which was developed to treat spinal injuries. In fact, this table—to which patients were strapped so that physicians could apply pressure and thereby correct the deformity—is still in use today and is considered by many to be the forerunner of the modern orthopedic table.

But one of the most intriguing facets of Hippocrates' medicine was his view that to preserve health or cure disease, it was necessary to understand the nature of the body *and* its environment. In other words, the body had to be treated as a whole, not simply a collection of unrelated parts. This view, in turn, was closely related to the concept of balance. While Hippocratic writings describe balance in differing ways, the basic view was that good health arose when forces in the body were in balance, while disease occurred when internal or external forces upset this balance. The physician's goal in treating patients, therefore, was to identify and correct any imbalance.

One of Hippocrates' best-known—but medically inaccurate—theories arose from the concept of balance. According to this theory, four humors, or fluids, circulate in the body: phlegm, bile, black bile, and blood. A person's state of health or disease arises from the degree of balance or imbalance among these fluids, along with their relation to the four seasons (winter, spring, summer, and autumn) and the four elements of nature (air, water, fire, and earth).

Although humoral theory is notably absent from modern text books of human pathophysiology, it can be argued that within this view lies the metaphysical roots of something deeper than modern medicine can fully explain.

* * *

Acknowledging Hippocrates' invitation to discuss his philosophy, Anaxagoras nodded silently and picked up a stick. Slowly and deliberately he began to speak, sketching out his thoughts in the dirt with a series of circles and lines…

"Things in the one universe are not divided from each other, nor yet are they cut off…" He paused to see Hippocrates was following along.

Indeed he was.

"Thus also," continued the philosopher, "all things would be in everything…and all things would include a portion of everything… Nothing could be separated, nor yet could it come into being of itself, but as they were in the beginning so they are now, all things together…"

Milestone #6 Where the two worlds meet: a holistic approach to medicine

> "It is necessary for a physician to know about nature, and be very eager to know, if he is going to perform any of his duties… what man is in relation to what he eats and drinks, and in relation to his habits generally, and what will be the effect of each upon each individual."
>
> —*Hippocrates,* Ancient Medicine, *420-350 BC*

It is not too great a leap to connect the philosophy of Anaxagoras with the holistic views that underlie much of Hippocratic medicine. According to some accounts, it was not long after he met with Anaxagoras in the ancient city of Miletus and learned of the philosopher's theory of matter and infinity that Hippocrates developed his view that human health cannot be separated from the natural surroundings. Whether or not the story is true, it points to a fundamental insight that forms the core of Hippocratic medicine. It can be found in his specific prescriptions for disease, as well as his general theories on medicine and on staying healthy. It points to the importance of the inner world, a person's own body or "constitution," and the outer world, the environment. In so doing, it points also to a place where the two worlds meet.

And where do the two worlds meet? From the perspective of patients and the extent to which they have any control over their health, there are at least three places where the internal (their bodies) meets the external (the outside world): food (diet), physical movement (exercise), and air (breathing). Hippocrates frequently emphasizes all of these factors and more in discussing his holistic view of medicine. And of course, regardless of which factor he is discussing, the overall goal of good health is to use these factors to maintain or restore balance.

For example, regarding diet and exercise, Hippocrates advises in *Regimen I* that physicians must understand not only a patient's individual constitution, but also the role of diet and exercise in his or her life:

> "He who is intending to write correctly about human regimen must first acquire knowledge and discernment of the nature of man as a whole…and the power possessed by all the food and drink of our diet… [But] eating alone cannot keep a man healthy if he does not also take exercise. For food and exercise, while possessing opposite qualities, contribute mutually to produce health."

In other writings, Hippocrates regards diet as indistinguishable from other treatments of the time, including bleeding and drugs. For example, the book *Regimen* lists the various qualities of different foods, while *Ancient Medicine* discusses the innumerable "powers" of food.

Hippocrates also writes often about the importance of air and breathing. In *Breaths 4*, he notes that "All activities of mankind are intermittent, for life is full of changes, but breathing alone is continuous for all mortal creatures as they exhale and inhale." In another writing, he adds, "It is air that supplies intelligence… For the whole body participates in intelligence in proportion to its participation in air…. When man draws in breath, the air first reaches the brain, and so is dispersed into the rest of the body, having left in the brain its essence and whatever intelligence it possess."

Although Hippocrates' theories of the environment would strain the capacity of even twenty-first century technology to verify, the concepts nevertheless have an underlying ring of holistic truth. In addition to explaining that different seasons play a key role in health and disease, he also contends that different regions, warm and cold winds, the properties of water, and even the direction that a city faces are important considerations. In *Airs, Waters, and Places*, he writes:

> "When a physician arrives at an unfamiliar city, he should consider its situations relative to the winds and the risings of the sun... He must consider as thoroughly as possible also the nature of the waters, whether the inhabitants drink water that is marshy and soft, or hard from high and rocky ground, or brackish and costive."

Finally, it's important to note that despite all we have said of Hippocrates' rational approach to medicine and denunciation of supernatural forces as a cause of disease, he was no atheist. Whether out of respect for the family tradition of Asklepieion priests or from the same intuition that informed his other philosophies, Hippocrates also believed that a higher power was a necessary precondition for good health.

Thus, while few people today understand the full range of Hippocrates' contributions to medicine, we should not forget that he is an original proponent of a uniquely holistic approach to medicine. In fact, this holism included what we now think of as both western *and* eastern medicine, with its acknowledgement of the importance of:

1. Rationale thinking and natural causes
2. The individual nature of health and illness
3. The role of diet, exercise, and environment
4. The value of ethics and compassion
5. A respect for a higher power

Hippocrates for yesterday, today...and tomorrow

> "Patients are anonymous... Their recovery practiced in rooms similar to cockpits..."
>
> —*Orfanos, 2007*

Although the physical form of Hippocrates disappeared from this world some 23 centuries ago, the body of his work—the collective writings and teachings for which we credit him the "Invention of Medicine"—remain alive and well in the twenty-first century. Medical students continue to cite his oath, physicians and surgeons continue to praise his anatomical and clinical insights, and many others continue to be inspired by his insights.

And yet...

To those who see little or no connection between ancient medicine and modern medicine of the twenty-first century, some would ask that you take a harder look at where we are today and where we may be heading. In a recent medical conference held on the island of Rhodes, Greece, a physician's opening lecture reviewed the history and accomplishments of Hippocrates. He then noted that after the flowering of Greek and Roman medicine and the transfer of this knowledge to the west in the middle ages by Arab scholars, the face of medicine began to change. Over the next four centuries, from the Renaissance to the urbanization, industrialization, and molecularization of medicine in the nineteenth and twentieth centuries, the field of medicine shifted from an emphasis on the routine and compassionate care of individual patients to an increasing focus on technology, economics, and business-oriented administration.

"Patients have become anonymous," Constantin Orfanos noted in his 2006 address to the European Academy of Dermatology and Venereology. "Surgical interventions are procedures, to be honored as a brisk code number; emergencies and patient recoveries are practiced in rooms similar to cockpits for electronic cybernetics..."

To prevent the industrialization of medicine and its conversion into pure business, many now believe we need to look to the ancient past, to the healing tradition that arose long ago on a small island in the Aegean Sea. We might do well to revisit and reconsider the words and writings of a man whose practice of medicine was truly holistic, encompassing not only rationality and clinical observation, but ethics, compassion, and even belief in a higher force.

Hippocrates would surely not discount the extraordinary advances made in medicine over the past four centuries. Rather, he might advise that we temper our relentless progress with the same philosophy that led him to the breakthrough that made modern medicine possible. He might suggest that we look a little deeper, seek the same place that he discovered and shared with his followers—that place where the inner and outer worlds meet, where health and disease are so precariously balanced.

How Cholera Saved Civilization:

THE DISCOVERY OF SANITATION

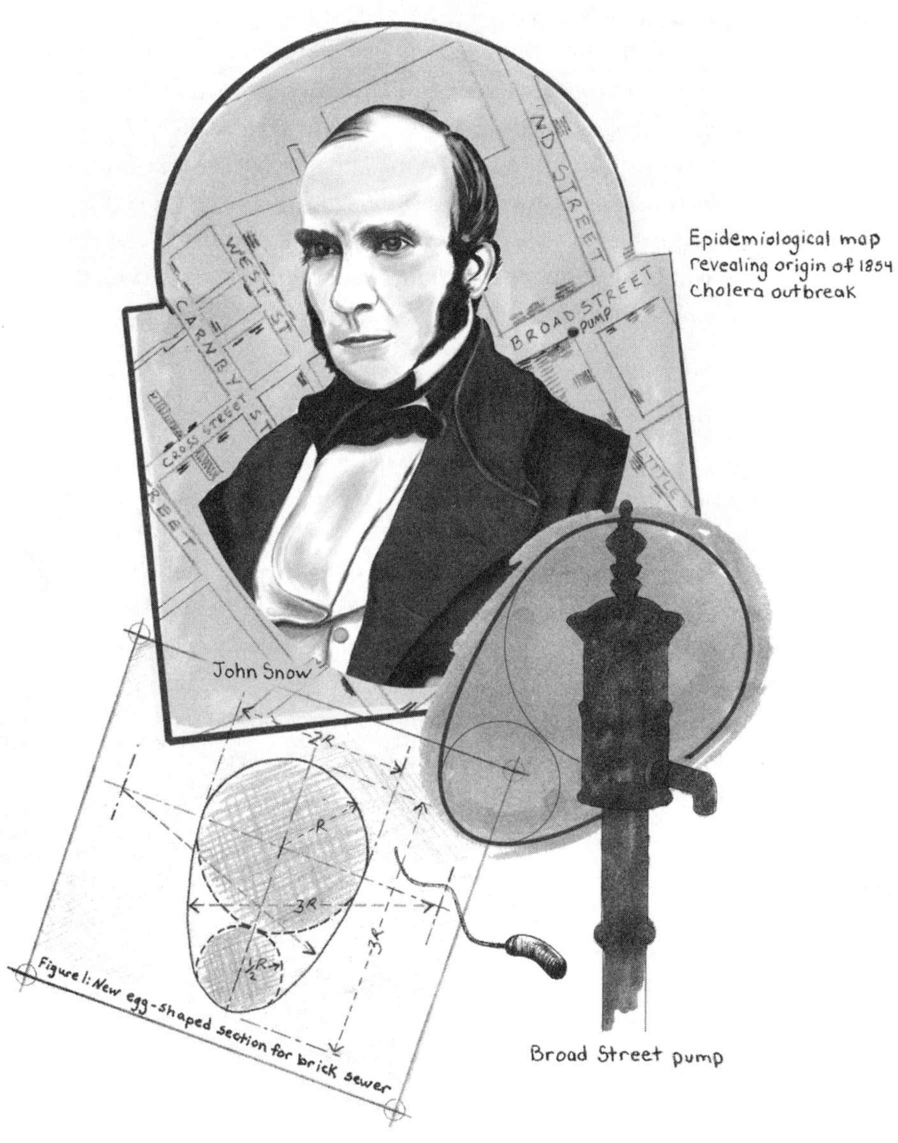

The largest delta in the world is a vast labyrinth of swampy waterways, tall grasses, mangrove forests, and brackish water. Formed from the Ganges and Brahmaputra rivers, it drains through 40,000 square miles along southern Bangladesh and a small corner of India before emptying into the Bay of Bengal. But the Ganges Delta is not merely large: It is also one of the most fertile regions in the world, churning with an enormous variety of life, from microscopic plankton to walking cat fish, from parrots, pythons, and crocodiles to the endangered Bengal tiger. In 1816, two somewhat less exotic life forms intermingled and formed a relationship that would quickly explode into deadly and global proportions. Within 15 years, it would kill thousands of people as it marched through India, parts of China and Russia, and into Europe. In October, 1831, it arrived on the northeast coast of England and quickly began to spread...

* * *

On December 25, 1832, John Barnes, an agricultural worker from a village 200 miles north of London, received what may be the worst Christmas present of all time. It was a box from his sister, who lived 22 miles away, in Leeds.

Barnes opened the box. It was not really a Christmas gift, and it's not clear whether Barnes was expecting what he found inside: They were clothes that had belonged to his sister, who had died two weeks earlier. As she had no children, the clothes had been dutifully packaged and sent to Barnes. Perhaps Barnes picked up the clothes and contemplated them for a few minutes, fondly recalling when he last saw her wearing them at a family gathering. Perhaps his wife held them up to herself to see if they would fit. In any case, before sitting down to dinner, both noticed something odd about the clothes: They hadn't been washed. By the time they had finished eating, nothing could stop what was to come next.

The next day, Barnes developed severe cramps and diarrhea. For two more days, the diarrhea was so severe that it did not let up. By the fourth day, Barnes was dead.

Shortly after Barnes had fallen ill, the same malady struck his wife. Her mother was notified, and she rushed to her daughter's assistance from a nearby village. Although Barnes' wife survived, her mother was not so lucky. After spending two days with her daughter and washing her linen, she left for home, just a few miles away. Somewhere along the road, she collapsed and was brought back to her village, where her husband and daughter were waiting.

Within two days, the mother, her husband, and her daughter were all dead.

In one sense, the deaths were no mystery. Local physicians recognized that the family had been stricken by cholera, the same disease that had broken out in England in the past 12 months. But in another sense, the deaths were nothing *but* a mystery. How could two families be stricken so suddenly and fatally when, until their deaths, no one in *either* of their villages had contracted the disease? Even the discovery that Barnes had received the unwashed clothes from his sister—who herself had died of cholera—shed little light on the mystery. After all, everyone knew at that time that cholera could not be transmitted in this way. With the discovery of pathogenic bacteria decades away, it was presumed—as it had been for centuries—that most disease was caused by inhaling *miasma*, the invisible particles released by decomposing organic matter, which might include anything from marshy waters and sodden ground, to garbage pits, open graves, and volcanic eruptions.

Yet one visionary doctor of the time *did* understand the significance of the Barnes story when he heard it. And though leading physicians would stubbornly reject the views of John Snow for the next half-century, he would ultimately not only be proven right, but would play a key role in the greatest medical breakthrough in history.

The Industrial Revolution: a new world of jobs, innovation—and spectacular filth

In 1832, the city of Leeds, like many cities in Europe and the United States, was beginning to experience everything both wonderful and terrible about the Industrial Revolution. In just a few decades, bucolic pastures, rolling hills, and woodlands had been transformed into brick-lined landscapes of textile mills and factories, their tall chimneys proudly puffing smoke into the newly urban skyline. But even as expanding industry generated new jobs and money, it also meant that more people—a *lot* more people—were flooding into cities to seek their fortunes. In just 30 years, the population of Leeds had more than doubled, creating housing problems never before seen on this scale: Thousands of workers and their families were literally being crammed into small rooms, overcrowded buildings, and jam-packed neighborhoods.

If it is unpleasant to imagine how such growth might strain a city's infrastructure, try imagining the impact before urban infrastructure had even been *invented*. For centuries prior to the Industrial Revolution, human waste from households and businesses was commonly disposed of in backyard pits, nearby alleys, and streets. From there, it was periodically removed by "night soil" workmen or scavengers who sold it as fertilizer or for consumption by pigs, cows, and other domestic animals. But with the explosive urban growth of the early 1800s, supply quickly exceeded demand, and streets, alleys, and cesspools were soon overloaded, clogged, and overflowing with waste.

According to one concerned official investigating sanitary conditions in Leeds at the time, "The surface of these streets is considerably elevated by accumulated ashes and filth… stagnant water and channels so offensive that they lie under the doorways of the uncomplaining poor, and privies so laden with excrementitious matter as to be unusable…" In many cases, overflowing cesspools rose up through the floorboards of a house or drained into nearby water cisterns and private wells for drinking water.

The public water supply was no better. One report found that the River Aire, a source of drinking water for many inhabitants of Leeds, was "charged with the contents of about 200 water closets [toilets], a great number of common drains, dead leeches and poultices from the infirmary, soap, blue and black dye, pig manure, old urine wash, and all sorts of decomposed animal and vegetable substances…"

Such was the setting in May, 1832, when cholera arrived in Leeds and claimed its first victim—a two-year-old child of a weaver who lived in "a small and dirty cul de sac inhabited by poor families." In six months—with no one understanding what it was or how it killed—cholera would take another 700 lives. Before subsiding later that year, more than 60,000 people in all of England would be dead of the disease. Although physicians and officials initiated frantic efforts to uncover and stop the culprit, over the next 35 years there would be three more epidemics that would claim more than 100,000 lives.

Nevertheless, well before the second epidemic, a prickly lawyer had begun to lay the groundwork for what would ultimately help end the ravaging epidemics and loss of human life. And although Edwin Chadwick was abrasive, brow-beating, and widely disliked, he, like John Snow, would play a key role in the greatest medical breakthrough in history.

* * *

When asked what the greatest medical advance is in the past two centuries, most people furrow their brows a moment and then give perfectly reasonable answers such as antibiotics, vaccines, x-rays, or even aspirin. When this question was recently posed to readers of the *British Medical Journal*, they came up with similar responses, along with some surprises, such as oral rehydration therapy, the iron bedstead, and salutogenesis. But when the *BMJ* tallied results from more than 11,000 readers across the globe, one medical advance beat out all others: Sanitation.

Sanitation broadly refers to the creation of a healthy environment through the provision of clean water, safe waste disposal, and other hygienic practices. Though it may not sound as technologically impressive as a polio vaccine or CAT scan, sanitation is arguably the most important of all medical breakthroughs because, once established, many diseases can be prevented in the first place. The principles of sanitation may seem obvious—most of us learn the basics of toilet training as toddlers—but at the dawn of the Industrial Age, the inability to provide sanitation on a large scale posed a genuine threat to the future of modern cities. It would take decades to even *conceive* of a reasonable solution, and decades more before the solution could be implemented.

While many individuals contributed to the development of sanitation, two people stand apart for their milestone insights and achievements. Although John Snow and Edwin Chadwick shared one thing in common—continuing battles with skeptical contemporaries—they could not have been more different in their personalities. Snow was described as "kindly in nature" and "always open and of sweet companionship," while the barrister Edwin Chadwick was a man "no one ever accused of having a heart" and possibly "the most hated man in England."

Nevertheless, from the 1830s to the 1850s, both were driven to solve the same mystery that had originated half-way across the world: What *was* killing tens of thousands of people and how could it be stopped?

* * *

It often begins with a sudden awakening at night—a roiling, squeezing urgency in the abdomen that sends one scrambling for the nearest toilet. Once there, relief is quickly replaced by the sickening depth of the purge. Though initially painless, the watery diarrhea is massive and alarming, the body discharging itself like fire hose. In a single day, more than five gallons of water may be lost. So intense is the purging that the intestinal lining is literally stripped and flushed away, the bits of tissue giving the diarrhea a characteristic "rice water" appearance. Before long, the first signs of dehydration—the final deadly blow—appear: muscle cramps, wrinkled and purplish blue skin, sunken eyes and pinched face, the voice reduced to a hoarse whisper. The disease strikes so suddenly that collapse and death can occur in hours. But even after death, the watery discharge itself continues to teem with life, seeking to infect others, wherever it may go...

Milestone #1 The first epidemic: a lesson from the depths of a coal mine

In the winter of 1831–1832, when John Snow was just 18 years old and barely underway with his medical apprenticeship, his surgeon-teacher sent him on unenviable mission: He was to go into the heart of a cholera epidemic, the Killingworth coal mine near Newcastle, to help the many miners who were suffering from a deadly disease for which there was no cure or treatment. Snow followed his instructions and, in the end, his tireless efforts to assist the miners were deemed a success. But perhaps more important, the experience left an indelible impression that would lead to his first milestone insight: If miasmas were truly the cause of cholera as everyone believed, how could the miners have contracted the disease while working in deep underground pits, where there were no sewers, swamps, or other miasmatic vapors to be inhaled?

As Snow later remarked in building his case that cholera was caused not by miasma but by poor sanitation:

> "The pits are without any privies, and the excrement of the workmen lies about almost everywhere so that the hands are liable to be soiled with it. The pitmen remain underground eight or nine hours at a time, and invariably take food down with them into the pits, which they eat with unwashed hands... Therefore, as soon as a case of cholera occurs among any of the pitmen, the disease has unusual facilities of spreading..."

After the first epidemic ended, Snow made his way to London, where he completed his medical training and pursued an entirely different medical field—the use of ether as an anesthetic during surgery. While he would eventually receive worldwide acclaim for this work—the subject of another chapter in this book—he never abandoned his interest in cholera. In fact, his research into the properties of inhaled gases only increased his doubts that cholera was caused by miasma. But with the first epidemic over, he lacked sufficient evidence to elaborate further on his theory that cholera was transmitted by the watery intestinal discharges of sick people.

Snow did not have to wait long for a new opportunity to gather more evidence. But would it be enough?

Milestone #2

Casting aside miasma to envision a new kind of killer

When the second outbreak of cholera hit London in 1848, 35-year-old Snow was mature enough to recognize the intersection of fate and opportunity when he saw it. As people began to die from an epidemic that would eventually claim another 55,000 lives, Snow began tracking the killer with a passion that bordered on obsession. Starting from square one, he learned that the first victim of this outbreak was a merchant seaman who had arrived in London from Hamburg by ship on September 22, 1848. The man had rented a room and died a short time later of cholera. Questioning the victim's physician, Snow learned that after the seaman died, a second person had the same room and died of cholera eight days later. Perhaps, Snow reasoned, something left behind by the first victim—for example, unwashed bed linen—had infected the second.

Snow continued his investigations and continued to turn up evidence that—contrary to the view of other medical authorities of the time—the disease was both contagious and could be transmitted by contaminated water. For example, he learned that in one section of London where two rows of houses faced each other, many of the residents in one row of houses developed cholera, while only one person in the other row became ill. Snow investigated and discovered that in the houses where people had been infected, "slops of dirty water, poured down by the inhabitants into a channel in front of the houses, got into the well from which they obtained their water..."

In another line of evidence, Snow noted that in every person stricken with cholera, just as he'd seen with the coal miners years before, the first symptoms were gastrointestinal—diarrhea, vomiting, and stomach pain. To Snow, the implication was clear: Whatever the "toxin" was, it must enter the body by swallowing contaminated food or water. If it were inhaled from some form of miasma, he reasoned, it would first enter the lungs and bloodstream, causing such symptoms as fever, chills, and headache.

Eventually, these and other observations enabled Snow to envision an invisible killer—an uncanny insight given that it would be decades before scientists would discover bacteria and viruses as causes of disease. But casting aside miasma theory, Snow concluded that cholera was caused by some kind of living agent that had "the property of reproducing its own kind" and "some sort of structure, mostly likely that of a cell." He further proposed that it "becomes multiplied and increased in quantity on the interior surface of the alimentary canal." Finally, he accounted for the incubation time before the first symptoms appeared by suggesting that "The period which intervenes between the time when [it] enters the system and the commencement of the illness which follows is a period of reproduction…."

In this way, Snow pushed the concept of germ theory further than anyone had prior to that time.

In 1849, hoping that his findings might lead to changes in policy and behavior that could end the outbreak, Snow published his views in a pamphlet, "On the Mode of Communication of Cholera." Yet despite his insights, Snow's colleagues were not impressed. While some grudgingly admitted that cholera *might* be transmitted from person to person "under favorable conditions," most contended that cholera was not contagious and, though related to poor sanitation, could not be transmitted by water.

Despite this setback, Snow did not give up. When the second epidemic subsided in 1849, he continued investigating other lines of evidence to support his theory. So far, he'd seen that in isolated outbreaks, such as in the coal mines and the Barnes family incident, cholera could be spread by poor hygiene and person-to-person contact. And he'd seen that in larger community outbreaks, cholera could be tracked to local wells that had been polluted by nearby cesspools. But to explain the massive scale of outbreaks that killed thousands of people, his aimed his sights on a new target: the public water supply.

It did not escape Snow's attention that at the time, the Thames River, a tidal river that flows through the center of London, served two contradictory public needs: sewage disposal and water supply. In fact, one of the city's sewage outflows emptied untreated into an area of the river where even sewage that had been carried away could be washed back during high tide. Investigating municipal records, Snow found that two major water suppliers—the Southwark and Vauxhall Company, and Lambeth Waterworks—pumped water from the Thames River to residents without filtration or treatment. However, in 1849, only one of these companies—Lambeth—took its water from an area of the river almost directly opposite the sewage outfall. Snow began collecting data, and his suspicions were soon confirmed: Communities who received their water from Lambeth had higher rates of cholera than those whose water came from Southwark and Vauxhall.

Snow was now on the brink of his two final milestones—just as London was about to suffer its third major outbreak of cholera.

Milestone #3 The invention of epidemiology and the disabling of a deadly pump

Although the third cholera epidemic began in 1853, it was not until August 31, 1854, that it would explode into the now-famous "Broad Street pump incident." In that incident, in less than two weeks, some 500 people who lived within 250 yards of the Golden Square area of Broad Street died of cholera. It was a mortality rate that, according to Snow, "equals any that was ever caused in this country, even by the plague."

But even before Snow would play his famed role in the Broad Street outbreak, he was investigating the Southwark and Vauxhall and Lambeth water companies for their possible roles in this epidemic. Since the 1849 epidemic, Lambeth had moved to an upstream location above the sewer outlets and was now providing cleaner water than Southwark and Vauxhall. Snow was intrigued when he discovered that the two companies—who supplied water to at least 300,000 people—piped water down the same streets, but to different houses. This enabled him to conduct an investigation "on the grandest scale." By determining which houses received which water supply, he could compare the numbers of people stricken with cholera against where they lived and whose water supply they received. Snow's epidemiological research did not let him down: During the first four weeks of the summer outbreak, the rates of cholera were *14 times* higher among those received Southwark and Vauxhall water compared to those who received the cleaner water from Lambeth. Once again, the evidence supported his theory that cholera could be transmitted by polluted water.

Snow was just beginning to sharpen his epidemiological tools. When the Broad Street epidemic broke out a few weeks later, on August 31, he immediately began new investigations. Over the course of several weeks, he visited numerous homes in the afflicted neighborhood and conducted interviews with sick people and their families. In this case, the water supply in question was from local wells, rather than the polluted Thames. Before long, Snow had identified all of the water pumps in the area, calculated their distances to the houses of the people who contracted cholera, and made a startling discovery: In one section, 73 of the 83 cholera deaths occurred in homes that were closer to a pump located on Broad Street than to any other pump, and 61 of the 73 victims had drunk water from that one pump.

This was strong evidence, and when Snow presented it to local officials, they agreed to shut down the Broad Street pump by removing its handle. But though this apparently ended the epidemic, it was not quite the victory Snow hoped for or is sometimes presented in popular accounts. For local officials, the idea that cholera had been transmitted by polluted water was still impossible to accept. Other factors could be found to explain why the outbreak ended and why water from the Broad Street pump might not have been the cause. For example, the outbreak may have ended not because the pump was disabled, but because the epidemic had already peaked, or because so many people fled the area when the outbreak began, and no one was left to be infected. But perhaps the most damning evidence against Snow's theory came when subsequent investigations found that the Broad Street pump water was *not* polluted.

Nevertheless, Snow remained convinced that the local outbreak was caused by contaminated water from the Broad Street pump. And in March 1855, in an extraordinary epilogue to the story, he would be vindicated by an unlikely hero…

* * *

Reverend Henry Whitehead was a deacon at St. Luke's Church who had no medical training and did not even believe Snow's theory that cholera could be transmitted by water. Nevertheless, impressed by Snow's investigations of the 1849 epidemic and compelled by the mystery of why the Broad Street outbreak had ended so quickly, Whitehead began his own investigations. Reviewing reports of the cholera deaths during the first week of the outbreak, Whitehead made a dramatic discovery: A five-month-old infant living at 40 Broad Street had died on Sept. 2, but her symptoms had begun several days earlier, *before* Aug. 31, when the massive outbreak began. Whitehead immediately recognized the significance of two key facts. The infant had to have been the *first* victim of the Broad Street outbreak, and she lived in a house at 40 Broad Street—which was directly *in front* of the Broad Street pump.

The rest of the story came together quickly. Whitehead interviewed the infant's mother, who recalled that at the time of her infant's illness, just prior to the full-scale outbreak, she had cleaned her child's diarrhea-soaked diapers in a pail of water. She then emptied the dirty water into a cesspool opening in front of the house. When inspectors were called in to examine the cesspool, they not only found that it was located less than three feet from the Broad Street well, but that the cesspool had been leaking steadily into the pump's well. With this discovery, Whitehead's original question was answered, and the mystery of the outbreak was solved: The first days of the outbreak coincided with the time when the diaper water was being emptied into the leaking cesspool; the epidemic subsided rapidly after the infant died and the diaper water was no longer being poured into the cesspool.

Yet although officials initially agreed with Whitehead and Snow that the new findings linked the contaminated pump water with the outbreak, they later rejected the evidence, convinced that some unknown miasmatic source must have been the cause.

* * *

When John Snow died of a stroke a few years later at the age of 45, the medical community still rejected his theory that cholera was caused by contaminated water. Yet it is satisfying to know that when the fourth and final outbreak of cholera hit London in 1866—ultimately claiming another 14,000 lives—it was Henry Whitehead who tracked the outbreak to a water company that had been supplying its customers with unfiltered water from a polluted river. And until he died in 1870, Whitehead kept a picture of Snow on his desk.

Physicians would continue to reject Snow's theory for several more decades. Finally, around the end of the nineteenth century, as bacterial theory began to displace the misconception of miasma, Snow began to be recognized for the achievements he made decades before the world was ready to believe him. Today, he is revered not only as the man who solved the mystery of cholera, but the father of modern epidemiology.

* * *

The true identity of cholera was first discovered in the very year that officials were rejecting John Snow's evidence for the cause of the Broad Street pump outbreak. In a separate outbreak in Florence, Italy, scientist Filippo Pacini had been studying intestinal tissue from cholera victims under a microscope and described what he saw in a paper published in 1854: tiny, rod-shaped organisms whose slight bend gave them a "comma-shape" and whose busy movement he described as "vibrio." Convinced the tiny organisms were responsible for causing cholera, Pacini published several more papers on the topic. Although John Snow never learned of Pacini's discovery, the two shared one thing in common: No one believed Pacini, either. His findings were ignored for the next 30 years. Even when Robert Koch, the founder of the science of bacteriology, "rediscovered" the bacteria in 1884, the best German scientists at the time rejected his conclusions in favor of a miasmatic explanation. Pacini eventually did receive credit—only a century late—when in 1965 the bacterium was officially named Vibrio Cholerae Pacini 1854.

Milestone #4

A new "Poor Law" raises hackles—and awareness

In the early 1830s, as John Snow was being duly praised for his first career milestone of helping cholera-stricken coal miners, a young lawyer named Edwin Chadwick also achieved his first career milestone—and was being duly despised. It's not surprising that Chadwick was hated for his role in creating the 1834 Poor Law Amendment Act: A key principle of the law was to make public relief *so* miserable to poor people that they would avoid it altogether. And from there, Chadwick's reputation only grew worse. Apart from being called an "oppressor of the poor" and possibly "the most hated man in England," he later became famous for battling officials, physicians, and engineers, for his overbearing and insensitive personality, and for being a man who did not so much converse with others as browbeat them into submission.

It's a good thing he turned out to be right.

In the end, Chadwick's bull-headed efforts would ultimately not only help *improve* the living conditions of the poor, but lead to the greatest medical breakthrough in history. The significance of Chadwick's first milestone was not the Poor Law itself, but what he achieved in his research to write the law. In fact, Chadwick did not oppose the poor so much as the deplorable conditions in which they lived. Like most people of the time, Chadwick realized that the growing unsanitary conditions in England's cities was somehow responsible for disease and the recent outbreak of cholera. Also, like most others, he was completely wrong about miasma causing cholera, to the extreme of publicly stating at one point, "All smell is disease."

Yet though technically wrong about the cause of cholera, Chadwick was right in principle, and in researching the Poor Law, he gathered an abundance of evidence linking unsanitary conditions with the living conditions of the poor. In fact, his documentation was so comprehensive—so much more thorough than anything done by his predecessors—that in designing the law he transformed policy analysis and drew widespread attention from his peers. Thus, even as he drew severe criticism for the Poor Law, Chadwick's research marked a key milestone that would soon lead to a reversal of fortune.

That reversal came in 1839. With sanitary conditions worsening, and in the wake of a two-year influenza epidemic, government officials decided it was time to take action. Impressed by Chadwick's thoroughness in writing the Poor Law, they asked him to report on the sanitary conditions and diseases in England and Wales and to provide recommendations for policy and technology solutions.

Chadwick accepted the new assignment, to the delight of the co-workers he was leaving behind: They'd found him impossible to work with.

Milestone #5
A grand report creates a wealth of ideas and a will to act

In 1842, after several years of research and writing, Chadwick released his report, *On the Sanitary Condition of the Labouring Population of Great Britain*. The fact that it was an immediate best-seller—selling more copies than any previous government publication—indicates how concerned people were about the sanitation problem. Compiled with the help of physicians and officials who described the conditions in their own towns and cities, the report presented an accurate image of the disease-causing filth plaguing many cities of England. At one point, referring to an epidemiological map of Leeds during the 1831–32 epidemic, Chadwick noted the clear link between unsanitary conditions and cholera. "In the badly cleansed and badly drained wards," he wrote, "[the cholera rate] is nearly double that which prevails in the better conditioned districts…"

But more than a roll-call of England's sanitary failures, the 1842 report was a milestone in several ways. First, it emphasized that the cause of poverty and disease, rather than a curse of God's will as many believed at the time, was due to environmental factors. Second, the report represented a culmination of a new public health movement that blamed poor sanitation on the industrial slums. Finally, and perhaps most impressive, it described Chadwick's groundbreaking ideas for an engineering and governmental solution—in short, the invention of modern sanitation.

A major part of Chadwick's grand vision was his proposal for an "arterial-venous" system. The first time anyone had viewed water and sewage as an interlocking problem, this "hydraulic" or "water carriage" system would pipe water into homes for the purpose of rinsing waste materials away through public sewers. It was a daring idea that proposed nothing less than a rebuilding of the urban infrastructure. It would require a city's terrain to be designed with proper street paving, sloping, and gutters so that the "self-cleaning" sewer pipes would remove sewage before decomposition would cause disease. Chadwick even proposed unique sewage pipes that were egg-shape in cross-section—rather than the usual circular design—to increase flow velocity and prevent solid deposits. Finally, rather than simply dumping sewage into the nearest river as many piecemeal systems of the time did, Chadwick wanted the waste to be directed to farms, where it would be recycled for agricultural use. In its sum, Chadwick's integrative sewer design was a first-of-its-kind; nothing like it existed in Europe or the United States.

Unfortunately, it was also far easier to describe than build. For although Chadwick also proposed new legal and administrative structures by which such systems could be financed and built, there was no existing model for how such a complex, citywide system could be implemented. At the same time, there were countless opportunities for various groups to argue over who should plan, build, finance, and maintain them. Nevertheless, after several years of legislative wrangling and haranguing by Chadwick and others, a solution finally emerged in 1848. Sort of.

Milestone #6

The long, slow birth of a public health revolution

The passage of the 1848 Public Health Act was considered to be the crowning point of Chadwick's work and a milestone in English public health. With this law, for the first time in history, the British government assumed responsibility for protecting the health of its citizens, for implementing the legal systems needed to guarantee sanitation.

But in reality, the law had numerous shortcomings that would not be solved for years. For example, despite passage of the Act, many guidelines were left to the discretion of local government. In some cases, Chadwick or his followers found themselves embarrassing, threatening, or harassing local governments into cleaning up their own filth. At that same time, those who were bold enough to attempt to build Chadwick's system often ran into technical difficulties that could not be resolved without compromising the grand plan. Thus, despite years of trying to make his sanitary system work—from abrasive arguments with engineers over technical details to accusations of the moral failings of opponents who stood in his way—Chadwick's grand vision ultimately turned out to be *too* ambitious.

Yet despite these setbacks, by the mid-1800s, Chadwick's work and vision began to manifest in positive ways. Though not as ambitious as the integrative system he had envisioned, urban sanitation systems that reflected his engineering and governmental ideas began to appear. And the early results were promising. According to one study of 12 large towns in Great Britain, death rates had dropped from 26 per 1,000 before sewage systems, to 17 per 1,000 after the systems were adopted.

What's more, by the 1860s and 1870s, the sanitary systems developed by Chadwick and other English engineers were having an international influence. In the 1840s, the first efforts to build sewers in large cities like New York and Boston had led to piecemeal, non-integrated systems with key design flaws. But by the time of the Civil War and into the 1870s, many U.S. cities had begun to implement "planned" systems based on what became known as "English sanitary reform." As one Massachusetts engineer of the time noted, "Our countrymen have seized upon the water-carriage system with great unanimity."

Back in England, the work of Chadwick and his followers finally culminated with the 1875 Public Health Act, the most comprehensive sanitary law in England to that date. Looking back on it now, the Public Health Act and proliferation of urban sanitation systems in the late 1800s could be traced back to three criteria that Chadwick had identified and championed as essential to modern sanitation: 1) recognition of the link between environment, sanitation, and health; 2) the need for a centralized administration to deliver and maintain sanitation services; and 3) a willingness to invest in the engineering and infrastructure needed to make such services possible.

* * *

One of the lessons of Chadwick's lifelong work is that as long as you're right, it's okay if it's for the wrong reasons. Throughout his life, Chadwick remained as adamant and misguided as his contemporaries in insisting that cholera was caused by miasma. Like many others, he was unimpressed by Koch's "rediscovery" of the *Vibrio cholerae* (*V. cholerae)* bacterium in 1883, and at one point he even argued that it was more important to remove foul smells from houses than to provide clean water. But even if technically wrong, you have to give him credit: He knew a bad thing when he saw—or rather—smelled it.

Today, Chadwick's achievements are viewed as a turning point in the history of modern sanitation. Set against the backdrop of widespread unsanitary conditions of the Industrial Revolution and 30 years of cholera epidemics, Chadwick raised awareness—and the bar—for the importance of sanitation to the health of a city and its inhabitants.

* * *

John Snow and Edwin Chadwick were men of great similarities and contrasts. Different in temperament and occupation, they were driven by a common enemy. Contrary in their views regarding the specific cause of cholera, both recognized the broader underlying problem to be a failure of human sanitation. In the end, Snow's epidemiological work and insights revealed to the world that contaminated water can spread serious gastrointestinal disease—what we now call the "fecal-oral route." And Chadwick's documentation linking poor sanitation and disease, along with his engineering and legislative innovations, helped make modern sanitation possible on an urban scale.

The work of Snow and Chadwick did not exactly overlap, but converged at a time when hundreds of thousands of people were vulnerable and terrified by an epidemic that could strike suddenly and wipe out entire families in days or even hours. In separate but additive ways, they helped "concentrate" the minds of an old world on the brink of a new age. Raising awareness, they helped nudge a reluctant humanity into a new phase of urban civilization, where modern sanitation would be essential to survival.

Cholera and the failure of sanitation: alive and well in the twenty-first century

In the twenty-first century, more than 150 years after it was first identified, *V. cholerae* remains alive, well, and deadly throughout much of the world in epidemic or endemic form. The good news is that today, with rapid oral rehydration and antibiotics, nearly all cholera deaths can be avoided. The bad news is that in many areas where cholera is a problem—including recent epidemics in Iraq, Rwanda, and Central and South America—treatment is not always available, and death rates remain as high as 50%.

While new vaccines offer better protection and fewer side effects than older versions, they remain limited by the difficulty of distributing them to the populations at risk—typically in developing or war-ravaged countries—and the need for frequent booster doses. What's more, even the best vaccines may be ineffective against the extraordinary numbers of cholera bacteria—as many as *100 million*—found in just one gram of watery diarrhea. Scientists point out that cholera will probably never be eliminated. Given that the natural habitat of V. *cholerae* coincides with the vast watery ecology of our planet, new epidemic strains are likely to always develop, evolve, and spread. Rather, scientists suggest that we learn to "get along" with V. *cholerae* by focusing on two basic goals: develop better ways to fight the causative organism and create better sanitation systems to prevent their spread.

Snow and Chadwick couldn't have put it better.

* * *

Perhaps what is most surprising about V. cholerae is that it is not one species, but a large family of ocean-loving bacteria—a family that is almost universally harmless. *Of the 200 known strains of V. cholerae, only two (called O1 and O139) possess the unique combination of genes needed to thrive in the intestines of human beings and produce their deadly toxin. One group of genes produces TCP, the substance that enables V. cholerae to colonize the intestinal lining; the other gene cluster, called CTX-ø, produces the lethal toxin that enters intestinal cells and convinces them to manically pump out every last drop of water until the human host dies. Curiously, while all 200 strains of V. cholerae live in brackish estuaries, O1 and O139 are the only two strains that contain the deadly cholera genes and are found in water polluted by* humans.

This raises an interesting question: Who contaminated whom?

Invisible Invaders: THE DISCOVERY OF GERMS AND HOW THEY CAUSE DISEASE 3

Shortly after 2:00 a.m. on a late-August night in 1797, Mrs. Blenkinsopp, midwife to the Westminster maternity hospital, hurried from the bed chamber, her pale face taut with anxiety. Three hours had now elapsed since she had delivered Mary's baby girl, and something had clearly gone wrong. She quickly found Mary's husband and gave him the alarming news: The placenta had still not been ejected; William must immediately call for help. The doctor arrived within the hour and, finding that the placenta was adhered internally, he began to operate.

But the surgery did not go well. The placenta could only be removed in pieces, and by the next morning, Mary had suffered significant blood loss and a night of "almost uninterrupted fainting fits." Nevertheless, as William later recalled, his beloved wife—whom he'd married just months earlier—mustered the strength to say she "would have died the preceding night but was determined not to leave me." After a small joke and weak smile, she added that she "had never known what bodily pain was before."

Mary had survived one crisis, but it was only the beginning. Several days later, as William and other family members kept watch with raised hopes for her recovery, Mary was suddenly overcome by fits of shivering so violent that "every muscle of her body trembled, her teeth chattered, and the bed shook under her." Although the rigors only lasted five minutes, Mary later told William that it had been "a struggle between life and death" and that she had been "more than once at the point of expiring."

Mary survived this crisis, too, raising hopes once again that she might recover. But over the next few days, she declined again, overwhelmed by such symptoms as high fever, an extraordinarily rapid pulse, and abdominal pain. Then, on the eighth morning after her delivery, just as William was once again giving up all hope, the surgeon woke him to report some extraordinary news: Mary was "surprisingly better."

Had Mary survived yet a third crisis? It certainly seemed so as, over the next two days, her shivering fits and other symptoms miraculously stopped. Indeed, by the tenth day after her delivery, the surgeon observed that Mary's "continuance was almost miraculous" and that it was "highly improper to give up all hope." Yet by this time, William's hopes had been raised and dashed once too often. And despite the remarkable cessation of symptoms, his sense of gloom proved prescient: On the eleventh morning following the birth of her daughter, Mary died from childbed fever.

* * *

When British author Mary Wollstonecraft died that late-summer morning in 1797 at the age of 38, the world lost more than a gifted philosopher, educator, and feminist. In addition to leaving a collection of writings that laid the foundation for the women's rights movements of the nineteenth and twentieth centuries and being the first woman to publicly advocate for women's suffrage and coequal education, Mary Wollstonecraft left one final, memorable gift to the world: The baby girl who survived the ordeal, named Mary in honor of the mother she never knew, grew up to be Mary Wollstonecraft Shelley, who in 1818, at the age of 19, wrote her well-known novel, *Frankenstein*.

Mary Wollstonecraft's death highlights the tragedy of a disease that was relatively common until the mid-1800s, usually fatal, and almost completely misunderstood by doctors. Although rare today, throughout history childbed fever, or puerperal fever, was the most common cause of death in women giving childbirth. As with Mary Wollstonecraft, it usually struck suddenly and unexpectedly shortly after childbirth, beginning with intense shivering fits, a pulse racing up to 160 beats per minute, and high fever. Lower abdominal pain was often so painful that the lightest touch or even the weight of bed sheets could trigger cries of agony. "I have seen some women," one obstetrician told his students in 1848, "who appeared to be awe-struck by the dreadful force of their distress." And in one final, cruel manifestation, the symptoms often stopped suddenly after days of suffering. But as family members rejoiced, experienced physicians recognized the ominous sign: The sudden absence of symptoms was an indication of advanced disease and usually meant that death was imminent.

But more than a historical footnote, childbed fever played a central role in a major turning point in medical history. When in 1847 Hungarian physician Ignaz Semmelweis discovered how to prevent it, he not only helped save countless women from agonizing deaths, he also took the first key step toward what is now regarded as one of the greatest breakthroughs in medicine: the discovery of germ theory.

The invisible "curiosity" that—finally—changed the world of medicine

Germ theory—the discovery that bacteria, viruses, and other microorganisms can cause disease—is something we all take for granted today. But until the late 1800s, the idea that germs could cause disease was so novel, even outlandish, that most physicians could not accept it without a monumental shift in thinking, a grudging surrender of long-held views, including miasma theory. In fact, vestiges of that nineteenth century struggle still remain with us today, as seen in the word "germ" itself. In the early 1800s, before microscopes were powerful enough to identify specific microbes, scientists broadly used "germ" when referring to these unseen and unknown microorganisms suspected of causing disease. Today, though we have long-known that germs are actually bacteria, viruses, and other pathogens, many of us—particularly ad copy writers hired to hawk kitchen and bathroom cleaners on TV—*still* use germ as a catch-all for any disease-causing microbe.

In any case, once "germ theory" had been proven by the end of the nineteenth century, it forever changed not only how doctors practiced medicine, but the very way we view, interact with—and often fear—the invisible world around us. The importance of germ theory was acknowledged in 2000, when *Life* magazine ranked it the sixth most important discovery in the past *1,000* years.

The initial reluctance to accept germ theory did not arise from any doubt that we live in a world surrounded and infused by invisibly small life forms. By the 1800s, the existence of microorganisms had been known for nearly two centuries. That key breakthrough had occurred in 1676, when Dutch lens grinder Antony van Leeuwenhoek, peering through his crude microscope, became the first human to see bacteria. On that April day, he reported with astonishment that he had seen a multitude of tiny "animalcules which... were incredibly small, indeed so small that...ten thousand of these living creatures would not be able to fill the volume of a small sand-grain."

But over the next two centuries, few scientists seriously considered that these invisible curiosities could cause disease. It was not until the 1800s that the evidence began to accumulate and, thanks to historic milestones by four key people—Ignaz Semmelweis, Louis Pasteur, Joseph Lister, and Robert Koch—germ theory was finally "proven." And the first of these milestone advances centered directly around the deadly mystery of childbed fever, the disease that not only took the life of Mary Wollstonecraft, but up to 500,000 other women in England and Wales during the eighteenth and nineteenth centuries.

Milestone #1

The tragic loss of a friend (and a brilliant gain of insight)

When Ignaz Semmelweis began his career in obstetrics at the Vienna General Hospital in 1846, he was just 28 years old and had every reason to be excited—and full of dread. The good news was that Vienna General Hospital was the largest of its kind in the world, and its affiliated Viennese School of Medicine was at its zenith. What's more, the maternity department had just been enlarged and split into two clinics, each capable of delivering up to 3,500 babies a year. But there was one terrible problem: The hospital was suffering a raging epidemic of childbed fever. While the death rate had been less than 1% in the 1820s, by 1841 it had increased by nearly 20 times. In other words, if you went to the Vienna General Hospital in 1841 to deliver your baby, you had about a one in six chance of not leaving the hospital alive.

By the end of 1846, when Semmelweis had completed his first year as a full assistant, he had seen more than 406 women die of childbed fever. By that time, numerous explanations for the high death rates had been proposed, both silly and serious. Semmelweis had considered, and ruled out, most of them, including theories that the deaths were due to: female modesty (in one clinic, babies were delivered by physicians, who were all males); bell-ringing priests (some thought that their marches through the wards after a death caused new cases by triggering fear); and other theories that did not fit the evidence, such as overcrowding, poor ventilation, and dietary errors.

But when Semmelweis conducted a statistical investigation comparing death rates between the two clinics, he immediately uncovered a compelling finding. In the five years after the maternity hospital had been split into two clinics, the death rate among women in the first clinic, where all deliveries were made by physicians, was three to five times *higher* than in the second clinic, where the deliveries were made by midwives. Yet, though suggestive, he could find no obvious reason for the discrepancy. As Semmelweis later wrote, the midwives who delivered babies in the second clinic "were no more skillful or conscientious in their duties" than the physicians who worked in the first clinic. His other investigations only confused the picture further. For example, death rates were actually *lower* in mothers who delivered their babies at home or even on the streets. As Semmelweis noted, "Everything was in question; everything seemed inexplicable; everything was doubtful. Only the large number of deaths was an unquestionable reality."

Then, in the spring of 1847, a milestone moment arrived in the form of personal tragedy. Returning to Vienna Hospital from a three-week vacation, Semmelweis was greeted with the "shattering" news that his much-admired friend Professor Jakob Kolletschka had died. Despite his grief, Semmelweis was intrigued by the cause of his friend's death: While performing an autopsy on a woman who had died from childbed fever, the professor's finger had been pricked by a medical student. The wound became infected and quickly spread through Kolletschka's body. During an autopsy, Semmelweis was struck by the widespread infection throughout Kolletschka's body and its similarity to what he had seen in women with childbed fever. "Day and night I was haunted by the image of Kolletschka's disease," he wrote, and the fact that "the disease from which he died was identical to that from which so many maternity patients died."

This dawning insight was remarkable in its implication. Until that time, childbed fever was, by definition, a disease that affected only women. The possibility that it had infected and killed a *man*, from a wound sustained during an autopsy of a patient who had died of the disease, led Semmelweis to a startling conclusion. "I was forced to admit," he wrote, "that if his disease was identical with the disease that killed so many maternity patients, then it must have originated from the same cause that brought it on in Kolletschka."

Although Semmelweis did not know what this "cause" was—he referred to the invisible culprit as "cadaver particles"—he had begun to solve the greater mystery. If childbed fever could be transmitted by "particles" from one person to another, it could explain the high death rates in the first clinic. Unlike midwives who delivered babies in the second clinic, physicians in the first clinic commonly performed autopsies on women who'd died of childbed fever and then went straight to the maternity wards where they conducted intimate examinations of women during labor. The answer to the mystery struck Semmelweis like a lightning bolt: It was the *physicians* who were transferring the infecting particles to the mothers, thus causing the higher death rates in the first clinic. "The cadaverous particles are introduced in the circulatory system of the patient," Semmelweis concluded, and "in this way maternity patients contract the same disease that was found in Kolletschka."

Although physicians *did* wash their hands after the autopsies, Semmelweis realized that soap and water was not sufficient—and thus he arrived at the next milestone.

Milestone #2

A simple solution: wash your hands and save a life

In mid-May, 1847, shortly after the death of his friend Kolletschka, Semmelweis announced a new practice in the first clinic: From now on, all physicians must wash their hands with a chlorine solution after performing autopsies and prior to examining pregnant mothers. Within the year, the new policy had a dramatic impact. Before implementing the hand-wash, the mortality rate in the first clinic had been about 12%, versus 3% for the second clinic. Just one year after the chlorine washings were instituted, the mortality rate had fallen to 1.27% in the first clinic, compared to 1.33% in the second clinic. For the first time in years, the death rates in the first clinic were actually *lower* than those in the second clinic.

But the reaction to Semmelweis' discovery underscores how far the medical world still had to go before accepting even this small step in the direction of germ theory. Although some colleagues supported his findings, many older conservative faculty rejected his ideas outright. For one thing, it contradicted the views held by most physicians that childbed fever, like most illness, was caused by many factors—from miasmatic vapors, to emotional trauma, to acts of God—and not some "particle." Furthermore, many physicians resented the implication that they were somehow "unclean" carriers of disease. And so, sadly, despite his discovery, Semmelweis' theory gained little acceptance. One problem was that he initially did little to promote his own findings. Although in 1861 he finally published a book on the cause and prevention of childbed fever, it was so rambling and repetitive that it made little impact.

From that point, Semmelweis' life turned progressively tragic as he succumbed to a serious brain disorder, possibly Alzheimer's disease. For example, in earlier years he had graciously described his sense of guilt and remorse for the unwitting role he and other physicians had played in transmitting childbed fever to so many women. "Only God knows the number of patients who went prematurely to their graves because of me… If I say this also of another physician, my intention is only to bring to consciousness a truth [that] must be made known to everyone concerned." But with his mental state deteriorating, all grace was lost as he began writing vicious letters to those who opposed his ideas. To one physician he wrote, "Your teaching, Herr Hofrath, is based on the dead bodies of women slaughtered through ignorance… If, sir, you continue to teach your students and midwives that puerperal fever is an ordinary disease, I proclaim you before God and the world to be an assassin…"

Eventually, Semmelweis was taken to a mental institution, where he died a short time later. Ironically, some contend that Semmelweis' final vitriolic attacks against his colleagues constituted a third key milestone: His abusive letters may have helped raise awareness years later, as other evidence for germ theory began to accumulate.

* * *

Although it would be another 15 years before those "cadaveric particles" would be identified as streptococci bacteria, Ignaz Semmelweis' insights are now recognized as a key first step in the development of germ theory. Despite having no understanding of the causative microbe, Semmelweis showed that a disease could have a single "necessary cause." In other words, while many physicians at the time believed that any disease could have multiple causes, Semmelweis showed that one specific factor, something in those cadaveric particles, *had* to be present for a person to develop childbed fever.

But it was only a first step. It would take the work of Louis Pasteur to push medical awareness to the next milestone: making a link between specific particles—microorganisms—and their effects on other living organisms.

Milestone #3 From fermentation to pasteurization: the germination of germ theory

As everyone knows, sometimes it can be downright impossible to get your hands on a rat or scorpion when you need one. But no worries, according to esteemed seventeenth-century alchemist and physician Jean-Baptiste van Helmont, who devised this recipe for making rats: "Cork up a pot containing wheat with a dirty shirt. After about 21 days a ferment coming from the dirty shirt combines with the effluvium from the wheat, the grains of which are turned into rats—not minute and puny, but vigorous and full of activity." Scorpions, Helmont assures us, are even easier: "Carve an indentation in a brick, fill it with crushed basil, and cover the brick with another. Expose the two bricks to sunlight and within a few days fumes from the basil, acting as a leavening agent, will have transformed the vegetable matter into veritable scorpions."

On the one hand, it's comforting to know that by the mid-1800s, most scientists would have joined us in laughing at such beliefs in spontaneous generation—the theory that living organisms can be created from nonliving matter. On the other hand, that laughter might trail off sooner than you would expect. Because as late as the 1850s, though no one seriously believed spontaneous generation could give rise to insects or animals, increasingly powerful microscopes had begun to prompt some scientists to rethink the issue when it came to the origin of organisms so small that 5,000 could fit in the span of the period at the end of this sentence.

Nevertheless, two vexing questions remained: Where *did* microorganisms come from, and did they have any relevance to the "real" world of plants, animals, and people? And so in 1858 the well-known French naturalist Felix Pouchet, attempting to answer the first question, resurrected the questionable notion of spontaneous generation, claiming that he had shown "beyond a shadow of a doubt" that it explained how microorganisms came into the world.

But French chemist Louis Pasteur, already admired for his work in chemistry and fermentation, didn't believe it for a moment and proceeded to devise a series of ingenious experiments that laid spontaneous generation in its grave forever. While Pasteur's classic experiments are still taught today in most biology classrooms, they comprise only a small part of a remarkable 25-year career. During this career, his contributions not only helped answer both questions—microbes come from other microbes, and they are *very* relevant to the real world—but raised the concept of germ theory from the mists of uncertainty to the brink of unquestioned reality.

A toast to yeast: a tiny critter gives birth to the liquor industry *and* a new theory of germs

To most of us, yeast is a powdery stuff that gives wine and beer their alcoholic pleasures and bread and muffins their ability to rise in a hot oven. Some of us also know that yeast is a single-celled microorganism that reproduces by producing small buds. We should be grateful for even these simple facts, as they represent the outcome of years of raging debate, controversy, and experimentation in the early 1800s. Even when scientists finally accepted that yeast was a living organism, it only set the stage for the next round of debates over whether it was truly responsible for fermentation.

An unsung hero of early microbiology, yeast was one of the first microbes to be studied scientifically because it is relatively large compared to bacteria. But often forgotten today is another major reason for its heroic stature: Thanks to the work of a multi-cellular scientist named Louis Pasteur, it played a central role in the development of germ theory.

It was an unlikely beginning. In 1854 Louis Pasteur was working as dean and professor of chemistry in Lille, a city in northern France, and had no particular in interest in yeast or alcoholic beverages. But when the father of one of his students asked if he would be willing to investigate some fermentation problems he was having with his beetroot distillery, Pasteur agreed. Examining the fermenting liquor under a microscope, Pasteur made an important discovery. Healthy, tiny globules in the fermentation juice were round, but when the fermentation became lactic (spoiled), the globules were elongated. Pasteur continued his studies, and by 1860 he had shown for the first time that yeasts were in fact responsible for alcoholic fermentation. With this discovery, Pasteur established the "germ theory" of fermentation. It was a major paradigm shift in thinking: the realization that a microscopic form of life was the foundation of the entire alcoholic beverage industry, that a single-celled microbe could indeed have very large effects.

In subsequent years, Pasteur extended his germ theory of fermentation into "diseases" of wine and beer, successfully showing that when alcoholic beverages went "bad," it was because *other* microorganisms were producing lactic acid. In addition to identifying the microbes, Pasteur devised a "cure" for this disease: Heating the liquor to 122 to 140 degrees F would kill the microbes and thereby prevent spoilage. The term for this process of partial sterilization remains well-known to us to this day, thanks to its ubiquitous presence on the packaging of many foods and beverages: pasteurization.

Pasteur's work in fermentation and diseases of wine was a major milestone in germ theory because of what it implied. As early as the 1860s, he was speculating about whether microorganisms could have similar effects in other areas of life. "Seeing that beer and wine undergo profound alterations because these liquids have given shelter to microscopic organisms, how can one help being obsessed by the thought that phenomena of the same kind can and must sometimes occur in humans and animals?"

Milestone #4 The "spontaneous generation of life" finally meets its death

It was while Pasteur was investigating fermentation that French naturalist Felix Pouchet ignited the scientific world with controversy and excitement by announcing he had "proven" spontaneous generation. Specifically, Pouchet claimed to have conducted experiments in which he had created microbes in a sterilized environment in which no germ "parents" were previously present. While many scientists discounted this claim, Pasteur's background in fermentation and his genius for designing clever experiments enabled him to take Pouchet head-on and disprove what many had believed was an unsolvable problem. In one classic experiment, Pasteur revealed the flaws in Pouchet's work by focusing on something so common that we tend to forget that it is as omnipresent as the air we breathe.

"Dust," Pasteur explained in a lecture describing his landmark experiment, "is a domestic enemy familiar to everyone. The air in this room is replete with dust motes [that] sometimes carry sickness or death in the form of typhus, cholera, yellow fever, and many other kinds..." Pasteur went on to explain how the germs that Pouchet claimed to have created through spontaneous generation instead arose from a combination of bad experimental technique and a dust-filled room. To prove his point, Pasteur described a simple experiment in which he poured a liquid nutrient into two glass flasks. One of the flasks had a straight, vertical neck that was directly open to the surrounding air and falling dust; the second flask had a long, curving horizontal neck that allowed air—but not dust—to enter. Pasteur then boiled the liquid in both flasks to kill any existing germs and set both flasks aside. When he checked the flasks a few days later, germs and mold had grown in the first, open flask, carried there by falling dust. However, the second flask, whose long neck prevented germ-laden dust from entering and contaminating the liquid, was germ-free.

Indeed, Pasteur explained, referring to the second flask, "It will remain completely unaltered not just for two days, or three or four, or even a month, a year, three years, or four! The liquid remains completely pure." In fact, having conducted similar experiments over the course of years with similar results, Pasteur rightfully claimed, "The doctrine of spontaneous generation will never recover from the mortal blow inflicted by this experiment."

Pasteur's 93-page paper describing his work, published in 1861, is now considered to be the final death blow to spontaneous generation. Equally important, his work set the stage for his next milestone. As he wrote at the time, "It is very desirable to carry these researches sufficiently far...for a serious inquiry into the origin of disease."

Milestone #5 The critical link: germs in the world of insects, animals, and people

For the next 20 years, Pasteur's work took a series of dramatic turns that, in addition to profoundly impacting health and medicine, collectively established the next milestone in germ theory. It began in the mid-1860s, when a mysterious disease was decimating the silkworm industry in Western Europe. When a chemist friend asked Pasteur if he would investigate the outbreak, Pasteur balked, pointing out that he knew nothing about silkworms. Nevertheless, intrigued by the challenge, Pasteur began studying the life history of silkworms and examining healthy and diseased silkworms under a microscope. Within five years, he had identified the specific disease involved, showed farmers how to prevent it, and thereby helped restore the silk industry to prosperity. But apart from the importance of his work to the silkworm industry, Pasteur had made another major step in the larger scheme of germ theory by entering the uncharted and complex world of infectious disease.

In the 1870s and 1880s, Pasteur extended his work in infectious diseases to animals and made several key discoveries that further contributed to germ theory. In 1877, he began to study anthrax, a disease that was killing as many as 20% of sheep in France. While other scientists had found a rod-shaped microbe in the blood of animals dying of the disease, Pasteur independently conducted his own work and, in 1881, stunned the world by announcing that he had created a vaccination that successfully prevented sheep from developing the disease. This major milestone in the development of vaccines (discussed in Chapter 6) added to the evidence that germ theory was real *and* relevant to diseases in animals.

But Pasteur had not yet finished his work with immunizations, and he soon began experiments to develop a vaccination for rabies, a disease that was relatively common at the time and invariably fatal. Although unable to isolate or identify the causative microbe—viruses were too small to be seen by microscopes at the time—he was convinced that some kind of germ was responsible. After hundreds of experiments, Pasteur created a vaccine that worked in animals. Then, in 1885, in a dramatic and risky act of desperation, the vaccine was used to successfully save the life of a young boy who had been bitten by a rabid dog. A crowning achievement in itself, Pasteur's vaccine extended germ theory to its culmination, showing its relevance to human diseases.

By the end of his career, Pasteur was a national and international hero, a chemist whose wide-ranging milestones had not only helped a diverse range of industries, but collectively provided powerful evidence for germ theory. Yet even with all his achievements, Pasteur's efforts alone still did not fully prove the concept of germ theory. A few more milestones would be needed, including a major development in 1865 by an English surgeon who was directly influenced by Pasteur's writings.

Milestone #6 Antiseptics to the rescue: Joseph Lister and the modern age of surgery

When Joseph Lister became Professor of Surgery at the University of Glasgow in 1860, even patients lucky enough to survive their operations had reason to fear for their lives. With postoperative infections an ever-present danger, the mortality rate for some procedures was as high as

66%. As one physician at the time noted, "A man laid on the operating table of one of our hospitals is exposed to more chance of death than the English soldier on the field at Waterloo." Unfortunately, efforts to solve this problem were thwarted by the belief that the "putrefaction" seen in postoperative infections was caused not by germs, but by *oxygen*. Many physicians believed that the festering of wounds resulted when oxygen from the surrounding air dissolved injured tissues and turned them into puss. And since it was impossible to prevent the oxygen in the air from entering wounds, many believed that it was impossible to prevent infections.

If Joseph Lister at any time believed this view, he began to change his opinion after reading the writings of Louis Pasteur. Two of Pasteur's ideas in particular stuck with Lister: that the "fermentation" of organic matter was due to living "germs"; and that microbes could only be reproduced from predecessor parents, rather than spontaneous generation. Given these observations, it occurred to Lister that when trying to prevent infections, it might make more sense to concern oneself with the germs—and not the oxygen—that could enter a wound. "If the wound could be treated with some substance which without doing serious mischief to the human tissues would kill the microbes already contained in it," he wrote, "putrefaction might be prevented however freely the air with its oxygen should enter."

After experimenting with several chemicals, Lister's milestone moment came on August 12, 1865, when he first used carbolic acid—"a compound which appears to exercise a peculiarly destructive influence upon low forms of life and hence is the most powerful antiseptic with which we are at present acquainted"—to treat an 11-year-old boy who sustained a compound fracture to his left leg after being run over by a horse-drawn cart. At that time, compound fractures had a high rate of infection and often required amputation. Lister splinted the boy's leg and periodically applied carbolic acid to the wound for the next six weeks. To his delight, the wound healed completely without infection. Lister later used carbolic acid to treat many other wounds, including abscesses and amputations. Eventually, he used it to disinfect incisions during surgery, as well as surgical instruments and the hands of his surgical staff.

Although Lister published his findings in 1867, as late as 1877 his work was met with skepticism by surgeons in London. Nevertheless, the value of his antiseptic techniques was eventually accepted, and today

Lister is often referred to as the Father of Antisepsis or the Father of Modern Surgery. In addition to the mouthwash named after him, Listerine, he has been honored by microbiologists who named the bacterial genus *Listeria* after him. Lister's discovery of aseptic surgery, for which he thanked Pasteur in a letter in 1874, has undoubtedly saved countless lives. But equally important, his discovery that germs play a key role in infection and can be eliminated through antiseptic treatment marked another key milestone in the development of germ theory.

Between the 1840s and 1860s, scientists had come a long way in gathering evidence for the role that germs played in disease. But up to that point, the evidence remained largely circumstantial. Even in the early 1870s, in the eyes of many, germ theory had still not been proven. But advocates and opponents agreed on one thing: To establish germ theory, someone needed to find a conclusive link between a *specific* microbe and a *specific* disease. The world would not have wait long before a young German doctor would conclusively show this link.

Milestone #7 One step closer: Robert Koch and the secret life of anthrax

In 1873, Robert Koch was a 30-year-old physician with a busy medical practice in a German farming district and all odds stacked against him. Despite being isolated from his peers, lacking access to libraries, and having no laboratory equipment other than a microscope given to him by his wife, he became interested in anthrax and set out to prove that it was caused by a specific microbe. By this time, a key suspect had been identified—a rod-shaped bacterium known as *Bacillus anthracis*—and Koch was hardly the first to study it. But no one yet had proven it was truly the cause of anthrax.

Koch's initial studies confirmed what others had found: Inoculating mice with blood from animals that had died of anthrax caused the mice to die from anthrax, while mice inoculated with blood from healthy animals did not develop the disease. But in 1874, he began investigating a deeper mystery that had been a key roadblock to proving that the bacteria caused anthrax: While sheep contracted anthrax when exposed to other sheep infected with the bacteria, why did some sheep also develop anthrax when exposed to nothing more than *soil*? After numerous experiments and painstaking work, Koch solved the mystery and opened a new window on the world of microbes and disease: In the course of its life cycle, anthrax can don a devilish disguise. During unfavorable conditions, as when they are cast off into surrounding soil, anthrax can form spores that enable them to withstand a lack of oxygen or water. When favorable conditions return, as when they are picked up from the soil and enter a living host, the spores are restored into deadly bacteria. Thus, the sheep who developed anthrax after seemingly being exposed to nothing more than soil had actually *also* been exposed to the anthrax spores.

Koch's milestone discovery of the anthrax life cycle and its role in causing disease brought him immediate fame. By establishing that *Bacillus anthracis* was the specific cause of anthrax, he nudged the medical community a giant step closer to accepting the concept of germ theory. But final proof and acceptance had to wait until he had solved the mystery of a disease that had long afflicted the human race. In the late-1800s, it infected nearly everyone living in a large European city and accounted for 12% of all deaths. Even today, it remains the most common cause of death due to an infectious agent and is responsible for 26% of avoidable deaths in developing countries.

Milestone #8

Sealing the deal: discovery of the cause of tuberculosis

When Koch first began studying tuberculosis, also known as consumption, the symptoms and outcome of this disease were well-known, even if its course was bafflingly unpredictable. A person who came down with TB might die within months, linger with the disease for years, or overcome it completely. When symptoms did appear, patients often initially had a dry cough, chest pains, and difficulty breathing. In later stages, the cough became more severe and was accompanied by periodic fevers, rapid pulse, and a ruddy complexion. In the final stages, patients became emaciated, with hollowed cheeks and eyes, throat ulcers that turned their speech into a hoarse whisper, and, as death became inevitable, a "graveyard cough." Many well-known people died from TB in the nineteenth century, including well-known artists such as Frederick Chopin, John Keats, Anton Chekov, Robert Louis Stevenson, and Emily Bronte.

Although scattered reports had previously suggested that tuberculosis was contagious, by the late 1800s many physicians still believed the disease was hereditary and caused by some kind of breakdown in the patient's lung cells. This breakdown, many believed, was quite possibly influenced by a person's mental and moral shortcomings, with no foreign invader involved. In the early 1880s, after being named director of the bacteriological laboratory at the Imperial Health Office in Berlin, Robert Koch set out to prove that, to the contrary, TB *was* caused by a microorganism.

It was no easy task, and in the course of his work Koch had to develop a number of new techniques, including a staining method that helped him distinguish the culprit microbe from surrounding tissues and a culture medium that enabled him to cultivate the slow-growing microorganism. But in 1882 Koch announced his discovery to the world: After successfully isolating, cultivating, and inoculating animals with the suspect microbe, he had found that tuberculosis was caused by the microbe, *Mycobacterium tuberculosis*. Using the term "bacilli" to refer to the rod-shaped bacteria, he concluded, "The bacilli present in the tuberculosis lesions do not only accompany tuberculosis, but rather cause it. These bacilli are the true agents of consumption."

Milestone #9 Guidelines for convicting a germ: Koch's four famous postulates

Koch's discovery of the bacterium that causes tuberculosis was the milestone that clinched the acceptance of germ theory. But more than that, the principles and techniques he used in his work on tuberculosis and other diseases helped him achieve one final milestone: a set of guidelines that scientists could apply when convicting *other* germs of causing disease. Called "Koch's Postulates," they stated that one could incriminate a germ by answering the following questions:

1. Is it a living organism with a unique form and structure?
2. Is it found in all cases of the disease?
3. Can it be cultivated and isolated in pure form outside the diseased animal?

4. Does it cause the same disease when a pure isolated culture of the microorganism is inoculated in another animal?

While Koch's discovery of the cause of tuberculosis eventually won him the Nobel Prize in physiology or medicine, his groundbreaking work in bacteriology continued after his work in tuberculosis. He eventually discovered (or technically, *re*discovered) the cause of cholera in 1883 and introduced public health measures that helped contain a cholera outbreak in Hamburg, Germany, in 1892. In addition, thanks to the microbiological techniques he developed, many of the co-workers he trained went on to discover other bacterial causes of disease. Although Koch later incorrectly claimed that he had discovered a treatment for tuberculosis, the extract that he developed—tuberculin—is still used today in a modified form to help diagnose tuberculosis.

Germ theory a century later: the surprises (and lessons) keep coming

Germ theory traveled a long and complex journey across the landscape of the nineteenth century. Interestingly, although it gradually grew in acceptance from milestone to milestone, the phrase "germ theory" did not even exist in the English medical literature until around 1870. But while the health benefits of germ theory soon became dramatically clear, often overlooked are some other key ways that it transformed the practice of medicine. For example, to many young physicians in the late 1800s, germ theory opened a new world of hope. Supplanting fickle theories of miasma and spontaneous generation, it implied that a cause—if not a cure—might be found for all diseases, which gave physicians a new authority in the eyes of their patients. As Nancy J. Tomes recently wrote in the *Journal of the History of Medicine*, by the late 1800s, physicians "began to inspire greater confidence not because they could suddenly cure infectious diseases, but because they seemed better able to *explain* and *prevent* them."

Germ theory also transformed physicians' understanding of how their *own* behavior could impact patient health. This new awareness was evident as early as 1887, when a physician at a medical meeting, hearing how another doctor had moved from an infected patient to several women in childbirth without washing his hands, angrily declared, "What amazes me is that a man of Dr. Baily's reputation as a teacher and practitioner of medicine would at this late date antagonize the germ theory of specific diseases... I trust no other member of this society will follow his implied example."

In fact, by the early 1900s, germ theory had literally changed the appearance of physicians: As part of a new code of cleanliness, young male physicians stopped growing the full beards so commonly worn by their older peers.

* * *

Today, despite its universal acceptance, germ theory continues to generate excitement, concern, controversy, and confusion throughout society. On the plus side, millions of lives continue to be saved thanks to our ability to identify, prevent, and treat diseases caused by microbes. Advances in technology have allowed us to see the smallest germs in existence, such as the rhinovirus, which causes the common cold and is so small that 500 *million* can fit on the head of a pin. The study of microbial diseases has taken us to the frontier of life itself, where scientists ponder whether viruses are actually "alive," and puzzle over how prion diseases such as Mad Cow disease can be infectious and deadly, even though the causative agent is clearly *not* alive.

More recently, our ability to decode the genome (the entire genetic make-up) of microbes has the led to new investigations that raise questions about the very nature of who we are. In 2007, the National Institutes of Health launched the "Human Microbiome Project," a project that will detail the genomes of hundreds of microbes that normally live in or on the human body. The idea that we even harbor a "microbiome"—the collective genome of all the microorganisms in our bodies—brings new meaning to germ theory. Given that there are 100 *trillion* microbes that inhabit the human body—10 times more than our *own* cells and comprising 100 times more genes than our *own* genes—where exactly is the dividing line between "us" and "them?" The fact that most of these microbes are essential to our good health—helping with normal body functions such as digestion, immunity, and metabolism—only mystifies the issue further.

In fact, since its discovery in the late 1800s, germ theory has opened a Pandora's box of anxiety that continues to mess with our minds. What is scarier than an omnipresent, invisible, and essentially infinite enemy that is able to cause terrible sickness and death? Who today has not thought twice before touching the doorknob or faucet in a public bathroom, shaking the hand of a stranger, or breathing in the stuffy air of a crowded elevator, bus, or airplane? While partly realistic, in susceptible people such concerns can develop into a full-blown anxiety disorders that literally dominate their lives. No wonder many of us think back wistfully on the pre-nineteenth century days of innocence, before germ theory stole away the bliss of our pre-hygienic ignorance.

For good or bad, the modern battle against germs has led to a proliferation of odd attire and habits throughout society, from the hair nets and surgeon's gloves worn by restaurant workers, to the antibacterial soaps, detergents, cutting boards, keyboards, and plastic toys now found in our homes. More recently, the battle against germs has led to a proliferation of squirt- and spray-bottle alcohol-based hand gels that have cropped up not only in doctor's offices and hospitals, but grocery stores, gas stations, purses, and back pockets. All of these measures—though criticized by some as potentially increasing bacterial resistance—point to a phobic undercurrent that runs through our lives, a hidden enemy against which we gladly aim the latest antiseptic weaponry in hopes of securing a little peace of mind.

Eliminating a few million unwanted guests: the answer is *still* at hand

Despite our efforts, it is not unreasonable to ask: Are we too aware or not aware enough? In fact, widespread failures of vigilance continue to cause huge numbers of people to fall sick and die every year, ironically, in the very places designed to make us well. According to a 2007 study by the Centers for Disease Control and Prevention (CDC), each year health-care-associated infections in American hospitals account for about 1.7 million infections and nearly 100,000 deaths. While many circumstances contribute to this high rate, one major factor is something Ignaz Semmelweis figured out long ago.

"If every caregiver would reliably practice simple hand hygiene when leaving the bedside of every patient and before touching the next patient," physician Donald Goldmann writes in a 2006 article in the *New England Journal of Medicine*, "there would be an immediate and profound reduction in the spread of resistant bacteria." In fact, studies have found that the number of bacteria on the hands of medical personnel range from 40,000 to as many as 5 *million*. While many of these are normal "resident" bacteria, others are "transient" microbes acquired by contact with patients and often the cause of healthcare-associated infections. At the same time, unlike resident bacteria that hide in deeper layer of the skin, these acquired microbes "are more amenable to removal by routine hand washing."

Although the CDC and other groups have been promoting hand-washing hygiene at least as far back as 1961, studies have found that healthcare worker compliance is "poor," often in the range of only 40–50%. This is unfortunate given that, according to the CDC, hand-washing or alcohol-based hand sanitizers have "been shown to terminate outbreaks in healthcare facilities, to reduce transmission of antimicrobial-resistant organisms, and reduce overall infection rates." Why the dismal rate of hand-washing? Various reasons given by healthcare workers include the irritation and dryness caused by frequent washing, the inconvenient location or shortage of sinks, being too busy, understaffing and overcrowding, lack of knowledge of the guidelines, and forgetfulness.

To his credit, Goldmann tries to be fair when discussing the negligence of healthcare workers. "The system is partly to blame," he writes, pointing out that hospitals must not overwork staff members so much that they don't have time for proper hygiene. He adds that hospitals need to educate caregivers, provide reliable access to alcohol-based antiseptics at the point of care, and implement a foolproof system for keeping dispensers filled and reliably functional. However, he warns, once a hospital has done its part, if caregivers continue to neglect hand hygiene, "accountability should matter."

When Ignaz Semmelweis made these points to his medical staff 160 years ago—with no knowledge of germs and only an intuitive awareness of their invisible presence—he helped save countless women from terrible suffering and deaths due to childbed fever. And though the medical community rewarded his efforts by ignoring him for the next 30 years, Semmelweis' milestone work ultimately nudged medicine forward on its first baby steps toward the discovery and acceptance of germ theory.

It is a "theory" that—no matter how compelling, well-established, and relevant to health, sickness, life, and death—many of us continue to grapple with to this day.

4 For the Relief of Unbearable Pain:

THE DISCOVERY OF ANESTHESIA

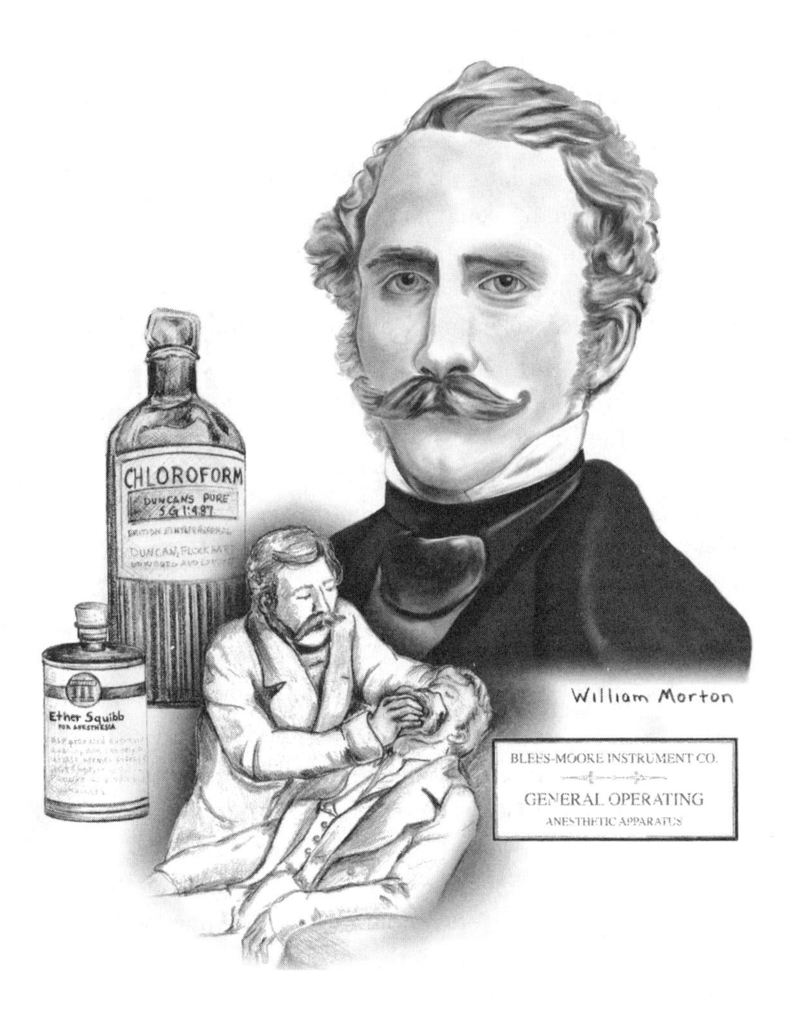

Even in today's world of high-tech medicine, where the traditional skills of medicine are increasingly lost to the convenience of digital sensors and gadgets, it's surprising how few physicians rue—or even remember—the lost art of nut-cracking.

It's a shame, because if you happened to have a knack for that sort of thing—the amount of force required to crack open various nuts based on the hardness of their shells—you might just have the skill needed to be anesthesiologist in the dark ages of medicine. For, as one ancient prescription directed: Place a wooden bowl on the patient's head, and knock him unconscious by striking the bowl with "sufficient strength to crack an almond, but leave the skull intact."

Or perhaps your talents lie in the art of delicate strangulation. In this forgotten method of anesthesia, practitioners asphyxiated their patients to the point of unconsciousness without, hopefully, also killing them. This method was used by the Assyrians prior to circumcising their children—undoubtedly without prior written consent—and was used in Italy as late as the 1600s.

Of course, many less traumatic methods have been used throughout history in an attempt to spare patients the pain of a surgeon's knife, including various opium preparations, the sleep-inducing seeds of henbane, the human-shaped mandrake root which, in addition to numbing pain, was said to emit a scream when pulled from the ground, and, the age-old favorite, alcohol.

Unfortunately, all early methods of anesthesia shared three key shortcomings: They didn't work well, or they killed you, or—in some cases—both. In fact, true anesthesia—defined as the ability to reliably and safely produce partial or complete loss of sensation, with or without loss of consciousness—was not officially "discovered" until 1846. This itself is painful to consider, given how many patients until that time suffered the most excruciating operations, from wrenching tooth extractions to gruesome amputations, with little or no pain relief. In fact, until the mid-nineteenth century, perhaps the only meaningful choice a patient had when selecting a surgeon was to ask how fast he was. Which is why you would have wanted someone like William Cheselden or Dominique-Jean Larrey at your operating table: The former, an English surgeon, could remove a kidney stone in 54 seconds. The latter, chief surgeon in Napoleon's army, could perform an amputation in as little as 15 seconds.

* * *

Sadly, for Fanny Burney, renowned nineteenth-century English novelist whose writings inspired Jane Austen, neither anesthesia nor surgical speed could save her from what must be among the most horrifying patient accounts of a major operation performed without anesthesia. On September 30, 1811, surgeons performed a full mastectomy to remove Burney's cancerous right breast, a procedure that lasted nearly four hours. Burney somehow managed to survive the ordeal and described it nine months later in a letter to her sister. As she recounted, her only "anesthesia" was a wine cordial (liqueur) and the benefit of not knowing about the operation until just two hours before it began. But even the short notice was of little help. "This, indeed, was a dreadful interval," she wrote. "Two hours thus spent seemed never-ending."

It's not hard to sympathize with Burney's dread when she first entered the room in her house that had been prepared for the operation. "The sight of the immense quantity of bandages, compresses, and sponges made me a little sick. I walked backwards and forwards till I quieted all emotions and became, by degrees, nearly stupid, torpid, without sentiment or consciousness, and thus I remained till the clock struck three."

Nor did her confidence improve when "seven men in black"—her physicians and their assistants—suddenly entered her home.

"I was now awakened from my stupor by a sort of indignation. Why so many and without my permission? But I could not utter a syllable… I began to tremble violently, more with distaste and horror of the preparations, even than of the pain."

A short time later, as Burney was guided onto the operating "mattress," she was given one final semblance of anesthesia: a linen handkerchief placed over her face to prevent her from seeing the proceedings. Unfortunately, even that failed its simple duty.

"It was transparent, and I saw through it that my bed was instantly surrounded by the seven men and my nurse. When, bright through the handkerchief, I saw the glitter of polished steel, I closed my eyes… A silence the most profound ensued, which lasted for some minutes, during which I imagine they took their orders by signs and made their examination. Oh, what a horrible suspension!"

The suspension was broken by more bad news when Burney, who had been expecting that only a limited amount of tissue would be removed, overheard their decision to remove her entire right breast. "I started up, threw off my veil, and cried out… I explained the nature of my sufferings, which all sprang from one point…"

But though the doctors listened "attentively," they responded with "utter silence." The veil was replaced, and Burney surrendered all resistance. The operation proceeded, as she recalled in vivid detail to her sister:

"When the dreadful steel was plunged into my breast—cutting through veins, arteries, flesh, nerves… I began a scream that lasted uninterruptedly during the whole time of the incision. I almost marvel that it doesn't still ring in my ear, so excruciating was the agony. When the instrument was withdrawn, the pain seemed undiminished, for the air that suddenly rushed into those delicate parts felt like a mass of minute but sharp and forked daggers that were tearing the edge of the wound."

And later, "When the instrument was withdrawn a second time, I concluded the operation was over—Oh no! Presently the terrible cutting was renewed and worse than ever… Oh Heaven! I then felt the knife tackling against the breast bone, scraping it!"

Burney recalled fainting twice during the operation, and finally, "When all was done, they lifted me up. My strength was so totally annihilated that I could not even sustain my hands and arms, which hung as if I had been lifeless, while my face, as the nurse has told me, was utterly colorless." She added, "For months I could not speak of this terrible business without nearly again going through it. Even now, nine months after it is over, I have a headache from going on with the account."

A painfully long wait: why it took 50 years for anesthesia to finally arrive

The good news is that Burney lived another 29 years after the operation. The bad news is that she need not have endured the horrors of surgery without anesthesia because in 1800—11 years *before* her operation—English scientist Humphry Davy had discovered something remarkable about a gas he had been experimenting with: "As nitrous oxide… appears capable of destroying physical pain," Davy wrote, "it may be probably used with advantage during surgical operations…"

This is the kind of prophetic statement that can drive historians crazy. If Davy had observed the "pain-destroying" properties of nitrous oxide as early as 1800—with others soon realizing that ether and chloroform had similar properties—why did it take nearly another 50 years for doctors to "officially" discover anesthesia? While controversies and debates abound, most historians generally believe that a mixture of religious, social, medical, and technical factors created a world in which many people in the early nineteenth century didn't want—or simply weren't ready for—anesthesia.

One clue to this mystery is seen in the word "pain" itself. Derived from the Greek word poine, or penalty, it implies that pain is a form punishment from God for some committed sin, whether or not the person was aware of it. Thus, for those who believed pain was a form of divine justice, attempts to alleviate it were fundamentally immoral and strongly resisted. The power of such thinking became dramatically clear when debates arose in the 1840s over the morality of giving anesthesia to women during childbirth. In addition, various social factors—including those perhaps best grouped under the heading "pointless bravado"—also played a role. Historians note that in almost all civilizations, the ability to endure pain has been viewed as a sign of nobility, virility, and character. And finally, some nineteenth-century physicians opposed pain prevention because they believed it served a necessary physiological function, the elimination of which might interfere with healing.

Yet as Burney's letter powerfully attests, many nineteenth-century patients facing the gleam of an approaching scalpel would have gladly welcomed the option of anesthesia. And most doctors would have gladly offered that option, if only out of self-interest. After all, nothing is so disruptive to the fine motor skills as a squirming, struggling, screaming patient. This was understood as far back as the fifth century BC, when the world's first physician explained his view on the matter. The role of the patient, Hippocrates wrote in a treatise on surgery, was "to accommodate the operator… and maintain the figure and position of the part operated on…" And, oh yes, as he comes at you with that scalpel, "Avoid sinking down, and shrinking from or turning away."

* * *

But to understand the factors that paradoxically delayed and led to discovery of anesthesia, we must look more deeply into the nature of anesthesia itself and its effect on human consciousness. Beginning in 1800, the discovery of medical anesthesia followed a quixotic, four-decade journey marked by a mixture of nobility and folly, curiosity and exhibitionism, courage and foolishness, callousness and compassion. And to begin that journey, one need look no further than the man who first observed and ignored the pain-killing potential of nitrous oxide. For it was Humphry Davy who, in the course of his scientific investigation of nitrous oxide, named the new gas Laughing Gas, inhaled up 20 quarts of it while seated in a sealed chamber and racing his pulse to 124 beats per minute, and who wrote of his experiences: "It made me dance about the laboratory as a madman and has kept my spirits in a glow ever since… The sensations were superior to any I ever experienced…inconceivably pleasurable…I seemed to be a sublime being, newly created and superior to other mortals…"

Milestone #1 From philanthropy to frivolity: the discovery (and dismissal) of nitrous oxide

Upon hearing that in 1798 an Englishman named Thomas Beddoes established a "Pneumatic Institution" in Bristol, England, many people today might imagine a group of scholars studying the design of jackhammers and tubeless rubber tires. In reality, the Pneumatic Institution for Inhalation Gas Therapy, as it was formally known, was a venture that pushed the frontiers of late eighteenth-century medical science. At the time, scientists had recently discovered that air was not a single substance, but a combination of gases. What's more, experiments by people like Joseph Priestly—who discovered nitrous oxide in 1772—revealed that different gases had different effects in the body. To enterprising people like Beddoes—well aware of the foul air now beginning to suffocate and sicken industrialized cities—the new science of gases created a market for health resorts and spas where people could be treated with various "therapeutic airs." Equally important, the Pneumatic Institute funded the *scientific study* of gases, and one of its most precocious and talented researchers was 20-year-old Humphry Davy.

Davy was put to work in the laboratory to investigate the effects of nitrous oxide, a job that included not only inhaling the gas himself, but inviting visitors to inhale and report how it made them feel. During one of his experiments, Davy noticed something peculiar about the gas: It relieved the pain he was experiencing from an erupting wisdom tooth. But though this discovery led to his famous observation about the potential of nitrous oxide to relieve surgical pain, Davy became sidetracked by other intriguing properties of the gas.

In his 1800 report titled "Researches, Chemical and Philosophical, Chiefly Concerning Nitrous Oxide or Dephlogisticated Nitrous Air, and its Respiration," Davy gave lengthy and vivid descriptions of these properties based on his own inhalations, including such entries as:

> "My visible impressions were dazzling... By degrees as the pleasurable sensations increased, I lost all connection with external things. Trains of vivid visible images rapidly passed through my mind and were connected with words in such a manner as to produce perceptions perfectly novel. I existed in a world of newly connected and newly modified ideas..."

When Davy asked volunteers who had inhaled nitrous oxide in his laboratory to write accounts of their experience, most reported being as amazed—and pleased—as Davy: "It is not easy to describe my sensations," a Mr. J. W. Tobin wrote. "They were superior to anything I ever before experienced. My senses were more alive to every surrounding impression. My mind was elevated to a most sublime height." Mr. James Thomson described "a thrilling sensation about the chest, highly pleasurable, which increased to such a degree as to induce a fit of involuntary laughter, which I in vain endeavored to repress..." And although some, like Mr. M. M. Coates, harbored suspicions that such reports were more due to an overactive imagination than pharmacologic effects, they quickly became converts: "I had no expectations of its influence on myself," Coates wrote, "but after a few seconds, I felt an immoderate flow of spirits and an irresistible propensity to violent laughter and dancing, which, being fully conscious of their irrational exhibition, I made great but ineffectual efforts to restrain..."

Attempting to better understand the effects of nitrous oxide on the body and mind, Davy even gave the gas to two paralyzed patients and asked how it made them feel. One reported, "I do not know how, but very queer," while the other said, "I felt like the sound of a harp." Davy thoughtfully wrote that the first patient probably had no analogous feeling with which to compare the sensations, while the second was able to compare it with a former experience with music.

As Davy continued to explore the visions and sensations produced by nitrous oxide, he contemplated their meaning with regard to philosophy and his interest in poetry. Thus, he formed a kind of club of artists—including the poets Robert Southey and Samuel Taylor Coleridge—with whom he could share the gas and discuss its effects on artistic sensibility.

Southey, after receiving the gas from Davy, raved that it gave him "a feeling of strength and an impulse to exert every muscle. For the remainder of the day it left me with increased hilarity and with my hearing, taste, and smell certainly more acute. I conceive this gas to be the atmosphere of Mohammed's Paradise." While Coleridge's response was more measured, he wrote to Davy that "My sensations were highly pleasurable… more unmingled pleasure than I had ever before experienced."

While this all might sound like the birth of a 1960s drug cult, it's important to understand that Davy's boss, Thomas Beddoes, was a physician and well-intentioned philanthropist whose goal in forming the Pneumatic Institution was to produce a revolution in medicine. By experimenting with various gases, he hoped to treat "excruciating diseases" as well as conditions where "languor and depression are scarce less intolerable than the most intense pain." One can't help admire the sincerity of his intentions—and thus the motivation behind Davy's experiments—when Beddoes wrote that he hoped to "diminish the sum of our painful sensations."

Yet despite such lofty ambitions, Davy's investigations of the euphoric effects of nitrous oxide ultimately distracted him from studying its potential for anesthesia. What's more, Davy eventually lost interest in nitrous oxide altogether: Within two years, he left the Institution to pursue other areas of scientific investigation. Although Davy later received acclaim for discovering the elements potassium, sodium, calcium, barium, magnesium, strontium, and chlorine, he never followed up on his observations of the "pain-destroying" effects of laughing gas. Indeed, within a few years, nitrous oxide was no longer being seriously studied at all. By 1812, one former enthusiast was warning in lectures that the gas "consumes, wastes, and destroys life in the same manner as oxygen wastes a taper, by making it burn too quick." Some historians even assert that nitrous oxide was "ridiculed into obscurity" by those who mocked the silly behavior of people under its influence.

And thus the first forays into anesthesia ran head-long into an ignominious and giggling dead end. Yet, putting aside the image of a dancing Humphry Davy careening madly about his laboratory, laughing gas should not be blindly condemned for its distractingly euphoric effects: It was those very properties that led to the next milestone.

Milestone #2 25 years of "jags" and "frolics" culminate in public humiliation—and hope

While medicine missed its chance to discover anesthesia in the early 1800s, the powers of nitrous oxide were not so quickly dismissed by other members of society. By the 1830s, reports began to surface that the recreational pleasures of inhaling nitrous oxide were being widely enjoyed—in both England and America—by virtually all strata of society, including children, students, entertainers, showmen, and physicians. About the same time, a new pleasure appeared on the scene, only to be similarly ignored by medicine and adored by the public: ether.

Unlike nitrous oxide, ether was not a recent laboratory discovery. It had been prepared nearly three hundred years earlier, around 1540, by the Swiss alchemist and physician Paracelsus. What's more, Paracelsus had observed that administering ether to chickens "quiets all suffering without harm, and relieves all pain." Nevertheless, it received little scientific attention until 1818, when Michael Faraday—famous for his work in electromagnetism—observed that inhaling ether vapors could produce profound lethargy and insensibility to pain. Unfortunately, taking a page from Davy's work with nitrous oxide, Faraday instead focused on the "exhilarating" properties of ether.

And so, by the 1830s, as physicians were condemning both nitrous oxide and ether as dangerous for medical practice, both gases were being embraced by the public for their exhilarating effects. According to one account published in 1835, "Some years ago… the lads of Philadelphia inhaled ether by way of sport… [causing] playfulness and sprightly movements…" Other accounts of the time refer to gatherings in which wandering lecturers and showmen invited people on-stage to inhale ether or nitrous oxide for the amusement of themselves and the audience. In fact, several pioneers of anesthesia claimed that it was their "ether frolics" during childhood that later inspired them to experiment with the gases for medical anesthesia.

Which brings us to what may be the first recorded "medical" use of ether for anesthesia. In 1839, William Clarke, like his fellow college students in Rochester, New York, had attended and participated in an ether frolic. Several years later, while a medical student at Vermont Medical

College, Clarke's experience triggered an idea. Under the supervision of his professor, he dripped some ether onto a towel and placed it over the face of a young woman who was about to have a tooth extracted. Unfortunately, whatever anesthetic benefit the woman gained from the ether was dismissed by Clarke's professor as an attack of hysteria, and Clarke was warned to abandon the further use of ether for such purposes. Thus, Clarke's milestone accomplishment received little attention, and he died unaware of his contribution to the discovery of anesthesia.

Around that same time, the recreational use of ether inspired another physician, who many believe should be credited as the true discoverer of anesthesia. Crawford Long had witnessed many nitrous oxide and ether "jags" while growing up in Philadelphia. Later, as a practicing physician in Georgia, he often inhaled ether with friends for its exhilarating effects. But apart from the euphoric effects, something else about ether caught Long's attention. As he later wrote, "I would frequently... discover bruised or painful spots on my person which I had no recollection of causing... I noticed my friends, while etherized, received falls and blows which I believed were sufficient to produce pain and they uniformly assured me that they did not feel the least pain from these accidents..." These observations were apparently on Long's mind in 1842 when he met with a Mr. James Venable, who had two small two tumors on the back of his neck. Venable was reluctant to undergo surgery due to his dread of the pain, but Long knew that the man was an enthusiast of inhaling ether. Recalling the pain-blunting effects he'd seen in himself and his friends, Long suggested that he give ether to Venable during the operation. Venable agreed, and on March 30, 1842, the operation was successfully and painlessly performed. But although Long went on to administer ether to many other patients, he neglected to publish his work until 1849—three years after another individual would receive credit for the discovery.

Not long after Long's first medical use of ether, another curious set of incidents led to a near miss in the discovery of anesthesia. In December, 1844, Horace Wells, a dentist living in Hartford, Connecticut, attended an exhibition in which a traveling showman, Gardner Colton, was demonstrating the effects of inhaling nitrous oxide. The next day, Colton put on a private demonstration for Wells and several others, during which a man who inhaled the gas proceeded to run wildly about the room, throwing himself against several couches and knocking them over, crashing to the floor, and severely bruising his knees and other parts of the body. Later, after the gas had worn off, the man marveled at the injuries he'd sustained and his lack of pain while under the influence of the gas, exclaiming, "Why, a person might get into a fight with several persons and not know when he was hurt!" Wells, suffering at the time from a painful wisdom tooth, was intrigued. He asked if Colton would give him the gas while another dentist removed the aching tooth. The following day, December 11, 1844, Colton administered nitrous oxide to Wells, the tooth was removed, and as the effects of the gas subsided, Wells exclaimed, "A new era in tooth pulling!"

But Wells' luck ran out when he attempted to introduce his discovery to the medical world. In January, 1845, he traveled to Boston to introduce anesthesia to the surgeons at Massachusetts General Hospital. One surgeon, John Warren, gave Wells an opportunity to administer nitrous oxide to a patient scheduled for a tooth extraction. Unfortunately, before a large audience of students and physicians, the gas "was by mistake withdrawn much too soon," and the patient groaned. Although the patient later testified the gas *had* lessened his pain, audience members called out "Humbug!" and Wells was laughed from the room.

And so, after decades of frolics and jags, dabbles and dismissals, along with the humiliations and unacknowledged successes of Clarke, Long, and Wells, a new milestone was finally imminent: The "official" discovery of anesthesia.

Milestone #3 Anesthesia at last: the discovery of Letheon (er, make that "ether")

When Horace Wells suffered his humiliating setback with nitrous oxide in front of a crowded room at Massachusetts General Hospital, it's not clear if one of the people in the audience—his former dental partner, William Morton—was among those who yelled out "Humbug!" In fact, Morton was probably as disappointed with the failure as Wells. Two years earlier, the two were working together on a new technique for making dentures that involved the painful removal of all the patient's teeth. Less than satisfied with their current anesthetic—a concoction of brandy, champagne, laudanum, and opium—both were on the look out for better ways to relieve their patients' pain and thus increase business. But though his former partner's demonstration of nitrous oxide had failed, it was around that time that Morton learned from an acquaintance, a professor of chemistry at Harvard Medical School, that ether had some interesting properties Morton might be interested in.

According to some accounts, Professor Charles Jackson had personally discovered these properties in 1841 after a vessel of ether exploded in his laboratory and he'd found his assistant anesthetized. After Jackson told Morton about these effects and provided information about how to prepare ether, Morton began his own personal studies. In a whirlwind series of trials possible only in an FDA-free world, Morton experimented on his dog, a fish, himself, his friends, and then, on September 30, 1846, a patient undergoing a tooth extraction. When the patient awoke and reported experiencing no pain, Morton quickly arranged for a public demonstration.

Two weeks later, on October 16, 1846—in what is now considered the decisive moment in the "discovery" of anesthesia—Morton entered the surgical amphitheater of Massachusetts General Hospital. Although late from making some final adjustments to the apparatus designed to deliver the gas, Morton administered the ether to Gilbert Abbot while surgeon John Warren removed a tumor from Abbott's neck. The demonstration was a success, and Dr. Warren, apparently familiar with a certain recent failure by Morton's partner, turned to the audience and announced, "Gentlemen, this is no humbug." The impact of the moment and its place in history was recognized by everyone present, including eminent surgeon Henry Bigelow, who said, "I have seen something today which will go around the world." Bigelow was right. The news was reported in the *Boston Daily Journal* the next day, and within months, the use of ether for anesthesia had spread to Europe.

Despite Morton's dramatic success, however, ether was almost immediately banned at Massachusetts General Hospital. Why? Morton had refused to tell the doctors exactly *what* he was administering. Claiming it was a secret remedy and under patent, he had added coloring and a fragrance to disguise the gas and called it "Letheon." But hospital officials were unimpressed and refused to use it further until Morton revealed its nature. Morton finally consented, and a few days later, Letheon—stripped of its coloring, scent, and name—made its reappearance at the hospital as plain old ether.

Although Morton spent the next two decades trying to claim credit and financial reward for discovering anesthesia, he ultimately failed, partly because Jackson and Wells were also fighting for the honor. Nevertheless, despite the contributions of various individuals over the preceding five decades—Davy, Clarke, Long, Wells, and Jackson—today Morton receives the widest recognition for being the first to demonstrate anesthesia in a way that profoundly changed the practice of medicine.

Milestone #4 Coming of age: a new anesthetic and a controversial new use

Despite the rapid and widespread use of ether, medical anesthesia had not yet truly arrived. One reason ether became popular so quickly after Morton's demonstration was that, whether by coincidence or fate, ether was blessed with a collection of almost too-good-to-be-true properties: It was easily prepared, markedly more potent than nitrous oxide, could be administered by simply pouring a few drops onto a cloth, and its effects were quickly reversible. What's more, ether was generally safe. Unlike nitrous oxide, it could be breathed in concentrations sufficient to produce anesthesia without risking asphyxiation. Finally, ether didn't depress heart rate or respiration and was nontoxic to tissues. Given the inexperience of those who first administered ether to patients—not to mention the clinically non-rigorous conditions seen at your average frolic and jag—nineteenth-century medicine could not have asked for a more ideal anesthetic.

In truth, however, ether was *not* perfect. Its limitations included the fact that it was inflammable, had an unpleasant odor, and it caused nausea and vomiting in some patients. As luck would have it, within a year after Morton's demonstration, a new anesthetic had been discovered—chloroform—and in a short time it almost completely replaced ether in the British Isles. The rapid acceptance of chloroform in England was probably due to several advantages it had over ether: It was nonexplosive, it had a less offensive odor and a speedier onset, and—perhaps most important—it was not discovered by that brash, young upstart, the United States.

Although chloroform had been first synthesized in 1831, it had not been tested in humans until someone suggested to Scottish obstetrician

James Simpson that he try it as an alternative to ether. Intrigued, Simpson did what any good researcher at the time would do—he brought some home and, on September 4, 1847, shared the potent gas with a group of friends at a dinner party. When Simpson later awoke on the floor, surrounded by his other unconscious guests, he became an ardent believer in the anesthetic properties of chloroform.

But Simpson did more than discover chloroform's anesthetic properties. Although the earlier discovery of ether had been rapidly accepted by medicine and society, its use remained highly controversial in one area: childbirth. This sensitivity was based on the religious view held by some that the pain of childbirth was God's just punishment for the sins of Adam and Eve. The outrage facing those who dared deny their moral punishment was violently seen in Simpson's own city of Edinburgh 250 years earlier: In 1591, Euphanie Macalyane had sought relief from her labor pains and, by order of the King of Scotland, was rewarded by being burned alive. Perhaps hoping to redeem the sins of his own ancestors, Simpson strongly advocated using anesthesia for painless childbirth, and on January 19, 1847, he became the first person to administer anesthesia—in this case, ether—to ease the delivery of a baby to a woman with a deformed pelvis. While Simpson faced angry opposition for his "satanic activities," he countered his critics by cleverly citing passages from the Bible, including suggesting that *God* was the first anesthesiologist: "…and the Lord God caused a deep sleep to fall upon Adam… and he took one of his ribs, and closed up the flesh instead thereof…"

Several months later, Fanny Longfellow, wife of the famous poet Henry Wadsworth Longfellow, became the first person in the United States to receive anesthesia during labor. In a letter she wrote afterwards, one can hear her mixed feelings of guilt, pride, anger, and simple gratitude for her pioneering role:

> "I am very sorry you all thought me so rash and naughty in trying the ether. Henry's faith gave me courage and I had heard such a thing had succeeded abroad, where the surgeons extend this great blessing much more boldly and universally than our timid doctors… I feel proud to be the pioneer to less suffering for poor, weak womankind… I am glad to have lived at the time of its coming...but it is sad that one's gratitude cannot be bestowed on worthier men than the joint discoverers, that is, men above quarreling over such a gift of God."

Milestone #5 From lint and gloves to modern pharmacology: the birth of a science

Although the use of ether after Morton's demonstration was rapid and widespread, anesthesia was not yet truly a science. To understand why, one need only read the words of a Professor Miller, who explained at the time that in the Royal Infirmary of Edinburgh, anesthesia was applied with "anything that will admit chloroform vapor to the mouth and nostrils." The "anything" Miller referred to included the nearest handy object, such as "a handkerchief, towel, piece of lint, nightcap, or a sponge" with, of course, special allowance for seasonal variations: "In the winter, the glove of a clerk or onlooker has been not infrequently pressed into service..." Miller added that dosing was less than an exact science: "The object is to produce insensibility as completely and as soon as we can, and there is no saying whether this is to be accomplished by fifty drops or five hundred."

One reason for this casual attitude was the perception that ether and chloroform were so safe. But, as one might expect, the increasing use of anesthesia was soon accompanied by more frequent deaths—sometimes suddenly and unexpectedly. In one 1847 medical report, a physician in Alabama wrote that he had been called to operate on a Negro slave suffering from tetanus and lockjaw. As the doctor heated his cautery to clean the wound, a dentist began administering ether to the Negro. But to the shock of everyone present, "In one minute, the patient was under its influence; in a quarter more he was dead—beyond all my efforts to produce artificial respiration or restore life. All present agreed that he died from inhaling the ether."

While such reports did not seem to concern most doctors, one man who became passionate, if not obsessed, with the use and safety of anesthesia was English physician John Snow. In 1846—two years before he would begin his milestone investigations into the outbreaks of cholera in London—Snow heard about the successful use of ether for anesthesia. Fascinated, he gave up his family practice and dedicated himself to study its chemical properties, preparation, administration, dosing, and effects.

Driven by his interest in safety, Snow investigated ether-related deaths, focusing on the role of overdosing and imprecise administration.

At a time when pharmacology was in its infancy, Snow impressively calculated the solubility of ether in blood, the relationship between solubility and potency, and even the role of room temperature in how much anesthetic entered a patient's body. Based on this work, Snow developed a device for vaporizing liquid anesthetics into gases, thereby creating a form of administration considerably more precise than, say, your average night cap or winter glove. Snow's improvement in the safety of anesthesia is clear from his detailed records where, among more than 800 cases in which he administered ether or chloroform to patients, he recorded only three deaths due to the use of anesthetics.

But perhaps the most influential and fascinating aspect of Snow's work was his clinical observations of patients as they underwent anesthesia. Prior to that time, most clinicians viewed anesthesia as a kind of "on/off" switch: Ether was administered, and the patient lost consciousness; surgery was performed, and the patient re-awoke. While it was obvious that patients experienced different stages of consciousness and awareness of pain, Snow was the first to seriously examine these stages and their relevance to safe, pain-free surgery. In his monograph, "On the Inhalation of the Vapour of Ether in Surgical Operations"—published in 1847 and now considered a classic in medicine *and* anesthesia—he not only provided guidelines for preparing and administering anesthesia, but identified five stages that are similar to the major stages of anesthesia recognized today:

- **Stage 1**—Patients begin to feel various changes but are still aware of where they are and can make voluntary movements.
- **Stage 2**—Patients still have some mental functions and voluntary movements, but they are "disordered."
- **Stage 3**—Patients become unconscious, losing mental functions and voluntary motion, though some muscular contractions may still occur.
- **Stage 4**—Patients are fully unconscious and immobile, with the only physical movements being the muscular motions of respiration.
- **Stage 5**—A dangerous final stage in which respiration is "difficult, feeble, or irregular," and "Death is imminent."

Providing details about these stages in a way that no clinician had before, Snow noted that patients generally passed from each stage to the next in one-minute intervals and that if the inhalation is discontinued after Stage 4, the patient will remain in that stage for one or two minutes before gradually passing back through Stage 3 (3 to 4 minutes), Stage 2 (5 minutes), and Stage 1 (10 to 15 minutes). He also wrote that surgery can be performed in Stage 3 "without producing any other effect than a distortion of features… and perhaps a slight moaning." In contrast, in Stage 4 patients "always remain perfectly passive under every kind of operation."

In describing how different types of patients react to anesthesia, Snow wrote that in passing from Stage 1 to 2, "hysterical females sometimes sob, laugh, or scream." He also found that a patient's memory of the experience usually only occurred in Stage 1 and that any reported feelings during this stage "are usually agreeable—often highly so." His guidelines included what patients should eat before anesthesia ("a sparing breakfast"), helping patients inhale ether ("The pungency of the vapor is often complained of at first… The patient must be encouraged to persevere," and a warning that in Stage 2, some patients may become excited and suddenly want to "talk, sing, laugh, or cry."

Snow's paper on ether was published in 1847, but before it had been widely distributed, James Simpson had introduced chloroform, and Snow soon began investigating the effects of this new anesthetic. Within several years, Snow had become an expert and was the favorite anesthesiologist for many of London's top surgeons. His fame peaked in 1853 and 1857, when he was asked to administer chloroform to Queen Victoria during her delivery of Prince Leopold and Princess Beatrice, respectively. "When the chloroform was commenced," Snow wrote, "Her Majesty expressed great relief…" And after the birth, "The Queen appeared very cheerful and well, expressing herself much gratified with the effect of the chloroform."

When Snow died in 1858, his research into the pharmacology and administration of anesthesia, along with his clinical experience and publications, had raised anesthesia to a science and made him the world's first true anesthesiologist. While the medical profession would not fully appreciate his work for many years, he had helped put the final exclamation point on one of the greatest breakthroughs in the history of medicine.

An underlying mystery: the link between losing consciousness and *raising* it

It's not hard to understand why some rank anesthesia as the top discovery in the history of medicine. After thousands of years of ineffective methods to prevent pain—from alcohol, to mandrake root, to a sharp knock on the head—the discovery of inhaled anesthetics was unlike anything previously seen or imagined. The ability to easily and completely remove patients from the awareness of pain despite the most drastic operations on the body, while allowing them to awaken minutes later with few or no after effects, transformed medicine and society. Patients were now willing to undergo more life-saving and life-improving procedures, and surgeons, freed from the hazards of a struggling patient, could perform more operations, while developing new techniques and life-saving treatments.

Yet as we saw in the early decades of frolics and jags, the discovery of anesthesia required a social transformation on several levels. Religious concerns had to be overcome, and physicians who believed pain was necessary for healing had to be enlightened. What's more, a new mindset had to arise in both physicians and patients that consciousness could be safely altered in this new and unimagined way. Most interesting, the pain-killing effects of anesthesia cannot, and perhaps should not, be separated from its effects on the mind. Looking back at the stories of Humphry Davy, Crawford Long, and Horace Wells, one can't help notice the irony that it was *euphoric* properties of these gases—and the physical injuries people sustained while enjoying them—that led to the discovery of their *anesthetic* properties.

Indeed, although medicine and society quickly put its focus on the anesthetic benefits of ether, thoughtful pioneers were immediately interested in the philosophical and metaphysical questions raised by its effects on the mind and body.

For example, John Snow, in the midst of his detailed scientific investigations, was intrigued by the comments from his patients as they awoke from anesthesia. "Some of the mental states… are highly interesting in a psychological view… The dreams often refer to early periods of life, and a great number of patients dream that they are traveling…" Snow added that even after the patient had recovered, "There is usually a degree of exhilaration, or some other altered state of feelings… The patient often expresses his gratitude to his surgeon in more ardent and glowing terms than he otherwise would do…"

Henry Bigelow, the surgeon who was present at Morton's milestone demonstration, also seemed curious about these effects when he wrote about several dental patients he observed as they were given ether. One patient, a 16-year-old girl, had a molar extracted. Although she had "flinched and frowned" when the tooth was removed, Bigelow reported that when she awoke, "She said she had been dreaming a pleasant dream and knew nothing of the operation." Another patient, "a stout boy of 12," had "required a good deal of encouragement" to inhale the ether. However, the youth was successfully anesthetized, two teeth were removed, and when he awoke, "He declared 'It was the best fun he ever saw,' avowed his intention to come there again, and insisted upon having another tooth extracted on the spot." A third patient had a back tooth removed, and when she awoke, "She exclaimed that 'It was beautiful.' She dreamed of being at home, and it seemed as if she had been gone a month."

Not surprisingly, some of the most descriptive accounts of how anesthesia affects the mind came from artists, thinkers, and philosophers of the time. Just one day after his wife became the first woman in the United States to receive anesthesia for childbirth, Henry Wadsworth Longfellow received ether for the removal of two teeth. He later wrote that after inhaling ether, "I burst into fits of laughter. Then my brain whirled round and I seemed to soar like a lark spirally into the air. I was conscious when he took the tooth out and cried out, as if from infinitely deep caverns, 'Stop,' but I could not control my muscles or make any resistance and out came the tooth without pain."

Even the earliest experimenters of nitrous oxide found that its effects raised fundamental questions about mental and sensory experiences and our limited ability to describe them. As one person who received the gas from Davy wrote, "We must either invent new terms to express these new and particular sensations, or attach new ideas to old ones, before we can communicate intelligibly with each other on the operation of this extraordinary gas."

Perhaps Davy was wise to seek the help of artists to put such experiences to words, for one of the best descriptions of how anesthesia awakens unexplored areas of consciousness came from American writer, naturalist, and philosopher Henry David Thoreau. On May 12, 1851, Thoreau received ether prior to a tooth extraction and later wrote, "I was convinced how far asunder a man could be separated from his senses.

You are told that it will make you unconscious, but no one can imagine what it is to be unconscious—how far removed from the state of consciousness and all that we call 'this world'—until he has experienced it.... It gives you experience of an interval as between one life and another, a greater space than you ever traveled. You are a sane mind without organs... You expand like a seed in the ground. You exist in your roots, like a tree in winter. If you have an inclination to travel, take the ether; you go beyond the furthest star..."

An evolving science: from "knocking out" a patient to administering precise molecular cocktails

The evolution of anesthetic drugs has traveled a long road since the pioneering discoveries of the mid-nineteenth century. Although nitrous oxide fell out of favor after Wells' embarrassing failure, in was revived in the 1860s for use in tooth extractions and later for some surgical procedures. Chloroform remained popular in Europe for a time, but was eventually found to have safety problems not seen with ether—including the potential to cause liver damage and cardiac arrhythmias—and its popularity soon declined. Of the three original inhaled anesthetic gases, only ether remained a standard general anesthetic until the early 1960s.

Throughout the early twentieth century, many new inhaled anesthetics were studied and introduced, including ethylene, divinyl ether, cyclopropane, and trichloroethylene, but all were limited by their flammability or toxic properties. By the 1950s, several inhaled anesthetics were made nonflammable by the addition of fluorine. While some were discontinued due to concerns about toxicity, inhaled anesthetics still in use include enflurane, isoflurane, sevoflurane, and desflurane.

Since the 1950s, anesthesia has advanced along many fronts, from the development of local, regional, and intravenous anesthesia, to technical advances in administering and monitoring anesthesia. But perhaps the most exciting advances are now coming from the frontiers of neuroscience. Although no one knows exactly how anesthetics work—any more than we understand the nature of consciousness—recent findings have provided clues into how anesthetics affect the nervous system, from their broad effects on consciousness and pain, to their microscopic and molecular actions on individual brain cells (neurons) in different areas of the brain and spinal cord.

At the broadest level, clinicians now understand that anesthesia is not simply a matter of "knocking out" a patient, but involves several key components, including: sedation (relaxation), hypnosis (unconsciousness), analgesia (lack of pain), amnesia, and immobility. In the 1990s, researchers discovered that anesthetics exert these multiple effects by acting on different parts of the nervous system. For example, the same anesthetic may cause hypnosis and amnesia by acting on neurons of the brain, while causing muscular immobility through its effects on neurons of the spinal cord. However, because no one anesthetic is ideal for producing all components of anesthesia, today's anesthesiologists usually select a combination of anesthetics to produce the desired effects while minimizing side effects.

* * *

Since the 1990s, researchers have uncovered even more surprising insights into how anesthetics work, opening new doors to better anesthetics for the future. For example, for years it was thought that all anesthetics acted on the same general target in the brain, broadly altering the neuron's outer membrane. However, researchers now know that there is no universal pathway that explains how all anesthetics work—or even how any *one* agent works. Rather, general anesthetics change the way that neurons "fire" (that is, transmit signals to each other) by altering microscopic openings on the surface of neurons, called ion channels. Because there are dozens of different types of ion channels, anesthetics can cause a variety of effects depending on which channels they act on. What's more, since the brain has literally billions of neurons and countless interconnections, the location of the affected neuron within the brain also plays a role. Researchers now know that major areas of the brain affected by anesthetics include the thalamus (relays signals to higher areas of the brain), hypothalamus (regulates many functions, including sleep), cortex (outer layer of the brain involved in thinking and conscious behavior), and hippocampus (involved in forming memories).

Even more exciting, neuroscientists have discovered in recent years that anesthetics produce their different effects by acting on highly specific "receptors." Receptors are tiny "gatekeeper" molecules on the surface of neurons that determine whether or not ion channels open (and hence whether the neuron will fire). Thus, when anesthetics attach to various receptors, they can affect whether or not the neuron will fire.

This is a key finding because there are many types of receptors, and researchers have learned that anesthetics may exert their unique effects—unconsciousness, sleepiness, analgesia, or amnesia—by binding to different receptors in different parts of the brain.

One particular receptor thought to play a key role in how anesthetics work is called $GABA_A$. Studies have shown that different anesthetics may cause different effects based on which *region* (subunit) of the $GABA_A$ receptor they attach to, where the $GABA_A$ receptor is located on the neuron, *and* where that neuron is located in the brain. With so many variables, it's clear that researchers have their work cut out for them trying to sort out the many pathways by which current and future anesthetics cause their effects.

Yet that's exactly what is most interesting and exciting about the future of anesthesia. As we gain a more precise understanding of how and where anesthetics act in the nervous system, it might be possible to develop drugs that target specific receptors and their subunits, while ignoring others, resulting in highly specific effects. In this way, customized anesthetics could be combined to create safer and more effective anesthesia. As anesthesiologist Beverley A. Orser noted in a recent article in *Scientific American*, the broad effects seen with current anesthetics are "unnecessary and undesirable." However, "With a cocktail of compounds, each of which produces only one desirable end point, the future version of anesthesia care could leave a patient conversant but pain-free while having a broken limb repaired…or a hip replaced."

* * *

And so today, 150 years after William Morton changed the practice of medicine by introducing ether to the practice of surgery, anesthesia continues to evolve and transform medicine. Equally intriguing, some believe that the insights gained from studying anesthetics might help uncover other secrets of the mind. As researchers noted in a recent article in *Nature Reviews*, "Anesthetic agents have been used to identify [neurons and pathways] involved in conscious perception [and the] mechanisms of sleep and arousal…" Results from ongoing studies "will probably provide further insights… which will be of great importance both for medicine and basic neuroscience."

Such insights could include not only how new anesthetics work, but the mysteries of human consciousness, from the nature of thoughts and dreams, to the sublime sensations and perceptions described by Humphry Davy more than 200 years ago. But wherever such studies lead us—whether deep within or “beyond the furthest star”—we should never lose sight of the life-altering benefits of anesthesia. As John Snow marveled in his 1847 paper, “The constant success with which ether is capable of being employed is one of its greatest advantages… and that the patient should not only be spared the pain, but also the anticipation of it… In most cases, the patients [can now] look forward to the operation merely as a time when they would get rid of a painful joint or some other troublesome disease…”

Snow’s statement speaks to the birth of a new science and a new awareness unknown before the nineteenth century, but deeply appreciated throughout the world ever since.

I'm Looking Through You: THE DISCOVERY OF X-RAYS

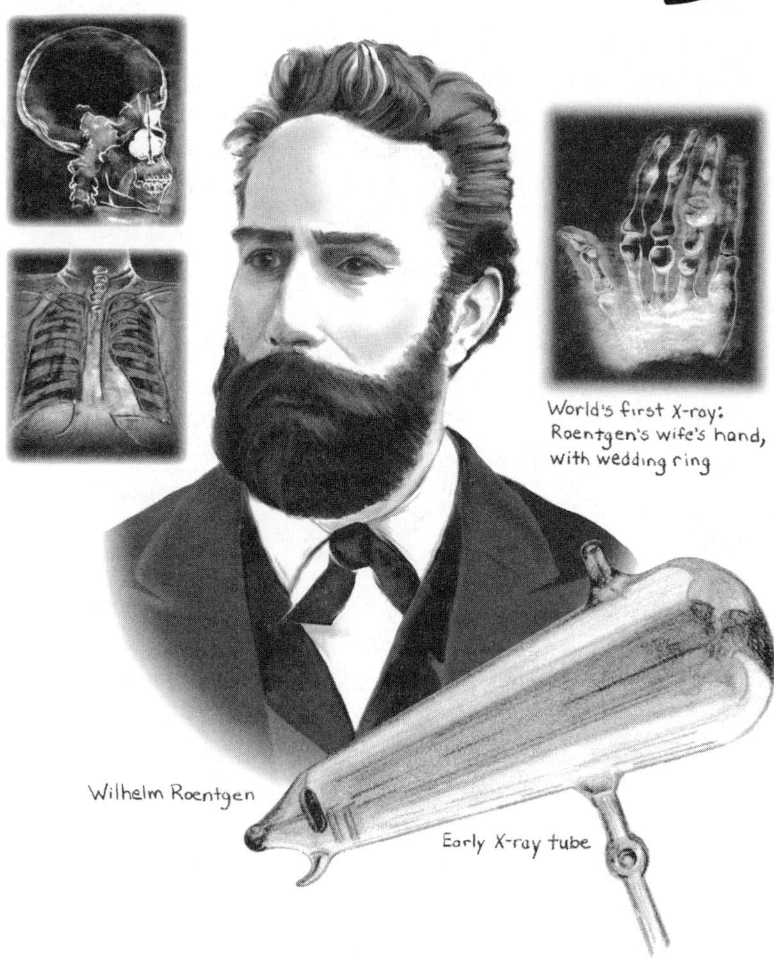

Mysteries, secrets, and revelations: four true stories...

Case 1: *Up until two days ago, the six-week-old boy had been healthy, active, and alert, but when his left thigh suddenly swelled up, his worried mother brought him to the emergency room. Responding to routine questions, she told doctors that the child had not been injured by rough play or an accident. What could explain the swollen leg—a tumor, blood clot, or perhaps infection? The mystery was solved with a single X-ray: Emerging from a shadowy black background, the ghostly white glow revealed a left thigh bone that had been cleanly snapped into two pieces. But an even darker secret was revealed by subsequent X-rays: The baby was also healing from fractures in his right forearm, right leg, and skull. Diagnosis in hand, treatment was clear. The boy was fitted with an orthopedic harness and placed—along with his two siblings—in a foster home to prevent further abuse.*

Case 2: *Jin Guangying, a 77-year-old grandmother in China, had been suffering from headaches for decades, some so severe that she would pound her head with her fist and babble incoherently. When her family finally borrowed enough money to take her to a doctor, X-rays of Guangying's head were unremarkable—a dull, grayish white landscape surrounded by a trace outline of her brain and facial bones. Except for* one *startling feature: There, nestled comfortably near the center of her brain, was the blazing white glow of a one-inch long bullet. Doctors removed the bullet in a four-hour operation and soon learned the full story. In 1943, during World War II, 13-year-old Guangying had been taking food to her father when she was shot by the invading Japanese army. She somehow survived the wound, and it was forgotten for the next 60 years—until a simple X-ray revealed both the mystery and the secret of her headaches.*

Case 3: *When the 62-year-old man arrived at the ER with stomach pain and an inability to eat or move his bowels, doctors were warned that he had a history of mental illness. But that hardly prepared them for what an X-ray of his chest and abdomen soon revealed. As doctors glanced over the billowy clouds of organs and ladder-like shadows of vertebrae, their gaze quickly fell on the enormous, bright-white sack in his lower abdomen. The shape did not correspond with any known anatomic feature—unless, of course, one's stomach happens to be packed with 350* *coins and assorted necklaces. As doctors learned during surgery, the 12 pounds of metal was sufficient to sink his stomach to its new location between the hips—explaining both the mystery of the patient's symptoms and the secret of a mental illness more severe than anyone had realized.*

Case 4: *Luo Cuifen, a 31-year-old woman from a rural province in China, suffered for years from depression, anxiety, and an inability to do physical labor. But it was not until she noticed blood in her urine that she went to the hospital for tests. What costly marvels of high-tech medicine would solve the mystery of her symptoms? Nothing more than a simple X-ray. There, glowing brightly among the shadows of her spine and pelvis, were the sharp outlines of 23 one-inch-long sewing needles lodged in her lungs, liver, bladder, and kidneys. As doctors prepared to operate, they revealed the dark secret behind this mystery. The needles were likely jabbed into Cuifen by her now-deceased grandparents when she was an infant—a failed attempt at infanticide in an area of rural China where baby girls are often killed because tradition does not allow them, unlike boys, to carry on the family name or support their parents in old age.*

Eerie and invasive: an invisible form of light that shocked and changed the world

Although these true stories—culled from recent medical journals and news reports—are unusual, they illustrate why X-rays continue to fascinate us 100 years after their discovery. In a glance, they can solve the deepest mysteries of pain and suffering, reveal unseen injuries and disease, and illuminate strategies for treatment. But as these stories also demonstrate, X-rays sometimes uncover even deeper secrets of human behavior—revelations of child abuse, wartime atrocities, mental illness, and the brutality of cultural shame. Today, we marvel—and sometimes fear—the humble X-ray for its ability to unveil truths that can, in a matter of seconds, change the course of a person's life.

As the world soon realized after their discovery in 1895, X-rays are a strange breed of science and magic. Looking at X-rays of our own bodies, we are confronted with an unsettling paradox: both the hard evidence of our sturdy inner workings *and* a skeletal reminder of our eventual demise and decomposition. Yet there is also magic in those images, the wonder that a trained eye can translate cloudy blurs and shadows into specific diseases and treatable injuries. This is a nice trick that, over the course of a century, has saved or improved millions of lives.

The mystique of X-rays is also hidden within its very name—the inscrutable "X" that suggests a power too unworldly to be pinned down by a "real" name. Indeed, X-rays are eerie because, in revealing our inner secrets, they *themselves* are secretive—invisible, unheard, and unfelt as they trespass our bodies at the speed of light.

* * *

Unlike most breakthroughs in medicine, X-rays were not discovered after a series of milestones in biology and health. Rather, their discovery was the result of decades of pioneering work in electricity and magnetism. Therefore, in this case we *begin* with the moment of discovery and then track the subsequent milestones that transformed X-rays into a breakthrough in medicine.

And those milestones are truly remarkable—from the shockwaves of amazement that shook the world after their discovery, to the countless applications that soon proved their unprecedented value in diagnostic medicine; from the discovery that they could treat cancer and other diseases, to the tragic realization that they could have dangerous, even deadly, effects. At the same time, X-rays helped trigger a paradigm shift in our understanding of reality itself. Indeed, they arrived at a time when scientists were grappling with the nature of the physical world—the structure of the atom and quantum physics—and for years no one knew exactly what X-rays were or how they could even exist. As Wilhelm Roentgen, the man who won the 1901 Nobel Prize in physics for discovering X-rays, once told an audience: Even after witnessing the rays pass through various objects, including his own hand, "I still believed I was the victim of deception."

But Roentgen soon became a believer, as did the rest of the world, once his first X-ray image was released to the public. That shadowy image—a photograph of his wife's hand clearly revealing bones, tissues, and the wedding ring on her finger—was unlike anything seen before. Almost immediately it set off a global firestorm of excitement, fear, and reckless speculation. As Roentgen later recalled, once the world saw that first X-ray, the secret was out and: "*All hell broke loose.*"

Milestone #1 How one man working alone at night discovered a remarkable "new kind of ray"

One Friday evening in November, 1895, a respected physicist in Germany began fooling around with something he had no business doing. Working alone in his laboratory, Wilhelm Roentgen began shooting electricity through a sealed, pear-shaped glass tube, causing its sides to emit an eerie fluorescent glow. It's not that Roentgen wasn't qualified: The 50-year-old Director of the Physical Institute at the University of Wurzburg had published more than 40 papers on various topics in physics. But he had shown no interest in such "electrical discharge" experiments until recently, when his curiosity had been aroused by an odd finding reported by another physicist.

For more than 30 years, physicists knew that firing high-voltage electrical discharges through a vacuum tube could cause the negative terminal in the tube—the cathode—to emit invisible "rays" that caused the tube to glow. They called these rays, logically enough, "cathode rays," though no one knew exactly what they were. Today, we know them to be electrons, the charged particles that orbit atoms and whose flow comprises electricity. But at the time, cathode rays were a mystery, and when in the early 1890s physicist Philipp Lenard discovered a new property—that cathode rays could actually pass *through* a small window of aluminum in the glass tube and travel a few inches outside—many scientists, including Roentgen, were intrigued.

On that historic evening, November 8, 1895, Roentgen was simply trying to repeat Lenard's experiment when two events—the products of curiosity and coincidence—led to a breakthrough discovery. First, he decided to cover the glass vacuum tube (called a Crookes tube) with light-proof cardboard and darken the room so he could better see the luminescent glow when the rays passed through the aluminum and outside the tube. Second, he happened to leave a small light-sensitive screen lying several feet away on a table.

Roentgen turned off the lights, fired up the Crookes tube, and watched for the faint glow to appear an inch or two outside the tube. Instead, something completely unexpected happened: An eerie yellow-green glow appeared in the dark *several feet* away, nowhere near the

Crookes tube. Roentgen scratched his head, checked his equipment, and repeated the discharges. The same strange glow appeared across the room. He turned on the light and immediately saw where the glow was coming from: It was the light-sensitive screen that happened to be lying nearby. Roentgen moved the screen around, fired up the Crookes tube, and checked and rechecked the glow until he could no longer doubt his eyes. Some kind of "rays" were coming out of the Crookes tube, striking the screen, and causing it to glow. What's more, they could not be cathode rays because to reach the screen, they had to travel at least six feet—25 *times* farther than the few inches that cathode rays were known to travel.

As Roentgen studied the rays late into that November evening and with feverish intensity over the next six weeks, he soon realized that the distance these invisible rays traveled was the *least* of their remarkable properties. For one thing, when they were beamed at the light-sensitive screen, the screen glowed even when the coated side was facing *away* from the rays. That meant the rays could pass through the *back* of the screen. Could they pass through other solid objects as well? In subsequent experiments, Roentgen found that the rays could easily pass through two packs of cards, blocks of wood, and even a 1,000-page book, before striking the screen and causing it to glow. On the other hand, dense materials, such as lead, blocked or partially blocked the rays, casting a shadow on the screen.

It was during these experiments that Roentgen made his final, astonishing discovery. At one point, while shooting the rays through an object to study its ability to stop the rays, he was startled to see cast upon the screen not only the shadow of his fingers holding the object, but within that shadow, the additional shapes of…his *own bones*. Roentgen had arrived at his milestone discovery. While he knew that the rays were absorbed in varying amounts based on the density of an object, this was a new twist: If the object *itself* consisted of different densities—such as the human body, with its bones, muscle, and fat—any rays passing through it would cast shadows of varying brightness on the screen, thus *revealing* those inner parts.

When Roentgen cast that first shadow of his own bones upon the screen, he simultaneously achieved two milestones: he had created the world's first X-ray *and* the first fluoroscope. But it was not until several weeks later, on December 22, 1895, that he created the world's first *permanent* X-ray image when he beamed the newly discovered rays through his wife's hand and onto a photographic plate.

Following his initial discovery, Roentgen worked alone and secretly for the next seven weeks. Sometimes sleeping in his laboratory and often skipping meals, he hardly hinted at his discovery to anyone other than perhaps his wife and a close friend or two. To one friend, he remarked with characteristic modesty, "I have discovered something interesting, but I do not know whether or not my observations are correct." During those weeks, Roentgen methodically explored the properties of these strange new rays, from the various materials they could penetrate, to whether, like other forms of light, they could be deflected by a prism or a magnetic field.

Finally, over the Christmas holidays, Roentgen wrote up his findings in a concise 10-page paper, titled "On a New Kind of Rays." In this paper, he used the term "X-rays" for the first time and reported—correctly—that the invisible rays were somehow generated when cathode rays struck the walls of the glass tube. On December 28, 1895, Roentgen sent his paper to the Physical-Medical Society of Wurzburg for publication in their *Proceedings*. A few days later he received reprints of the article, and on New Year's Day, 1896, he mailed 90 envelopes with copies of the article to physicists throughout Europe. In 12 of the envelopes, he included nine X-ray images that he had created. Most of the images showed the interiors of common objects, such as a compass and a set of weights in a box. But there was one image in particular—the image of his wife's skeletal hand bearing a ring—that caught the world's attention.

It took only three days for "All hell to break loose." At a dinner party on January 4, 1896, one of the recipients of Roentgen's article and X-ray images happened to show it to a guest from Prague, whose father happened to be the editor of *Die Presse*, Vienna's largest daily newspaper. Intrigued, the guest asked to borrow the images, took them home to his father, and the next morning, the story of Roentgen's discovery appeared on the front page of *Die Presse* under the headline, "*A sensational discovery.*" Within days, the story had been reported by newspapers across the world.

Milestone #2

A one-year firestorm and those "naughty, naughty" rays

It is almost impossible—no, it is *literally* impossible—to overstate the intensity and range of reactions by scientists and the public during 1896, the first year after Roentgen's discovery. From the moment Roentgen's shared his findings, even respected colleagues were stunned. One physicist to whom Roentgen sent the original reprint and photographs recalled, "I could not help thinking that I was reading a fairy tale… that one could print the bones of the living hand upon the photographic plate as if by magic…" A physician recalled that shortly after the first news reports came out, a colleague came to up to him at an event and excitedly began describing Roentgen's "peculiar" experiments. The physician scoffed and starting cracking jokes until the colleague became angry and left. But when the physician later met with a group of other doctors who were discussing the report, he read the article for himself and, "I must say that I was speechless."

Before long, there were few doubters. As the London *Standard* reported on January 7, "There is no joke or humbug in the matter. It is a serious discovery by a serious German Professor." And with acceptance, recognition of the implications quickly followed. On January 7, the *Frankfurter Zeitung* wrote, "If this discovery fulfills its promise, it constitutes an epoch-making result… destined to have interesting consequences along physical as well as medical lines." Later in January, *The Lancet* noted that the discovery "will produce quite a revolution in the present methods of examining the interior of the human body." And on February 1, the lead article in the *British Medical Journal* stated that "The photography of hidden structures is a feat sensational enough and likely to stimulate even the uneducated imagination."

In those first few weeks, many scientists reacted exactly as you might expect: They raced out and bought their own Crookes tubes and equipment—which at the time cost less than $20—to see if they could create their own X-rays. In fact, so many people did this in the first month that on February 12, 1896, the *Electrical Engineer* wrote, "It is safe to say that there is probably no one possessed of a vacuum tube and induction coil who has not undertaken to repeat Professor Roentgen's experiments." One week later, *Electrical World* reported, "All the Crookes tubes in Philadelphia have been purchased…" Telegraph wires, too, were abuzz with scientists seeking advice. When one physician in Chicago wired inventor Thomas Edison for technical advice, Edison wired back the same day, "THING IS TOO NEW TO GIVE DEFINITE DIRECTIONS. IT WILL REQUIRE TWO OR THREE MORE DAYS EXPERIMENTING..."

As the news spread and became the hot topic of the day, some couldn't restrain their cynicism with the hullabaloo. In March, the English *Pall Mall Gazette* noted, "We are sick of the Roentgen rays... It is now said... that you can see other people's bones with the naked eye... On the revolting indecency of this there is no need to dwell." And on February 22, 1896, the editor of the *Medical News* wrote, "It is questionable how much help can be obtained by such crude and blurred shadow pictures..."

But for many scientists, there was little question of the significance of X-rays. On January 23, 1896, Roentgen gave one of his few public lectures on his discovery to a large group that included members of the Physical-Medical Society of Wurzburg, university professors, high-ranking city officials, and students. Roentgen was greeted with a "storm" of applause and interrupted repeatedly during his talk by more applause. Near the end, he summoned famous anatomist Rudolph von Kolliker from the audience and offered to make an X-ray of his hand on the spot. The X-ray was made, and when the image was held up to the room, the audience again burst into applause. Von Kolliker then praised Roentgen and led the crowd in calling three cheers for the professor. When von Kolliker concluded by suggesting the rays be named after Roentgen, the room again thundered with applause.

Perhaps the best evidence for the overwhelming interest in Roentgen's discovery during that first year is seen in a simple statistic: By the end of 1896, more than 50 books and 1,000 papers about X-rays had been published worldwide.

As for the general public, the response was equally enthusiastic—but far more charged by irrational fears, nervous humor, and shameless profiteering. One of the biggest initial misunderstandings was the belief X-rays were just another form of photography. Many early cartoons made great fun of this misperception, such as joke in the April 27, 1896, issue of *Life* magazine. A photographer is preparing to take a picture of a woman and asks her if she would like it "with or without." She answers, "With or without *what*?" To which he replies, "The bones."

From such misunderstandings came genuine fears that shady individuals, driven by prurient desires, would take their X-ray "cameras" to the streets and snap revealing photographs of innocent passersby. And so, within weeks of the discovery, one London company thoughtfully advertised the sale of "X-ray-proof underclothing—especially made for the sensitive woman." In a similar vein of misunderstanding, Edison was undoubtedly puzzled when he received two odd requests in the mail. In one, the libidinous individual had sent a set of opera glasses, asking Edison to "fit them with X-rays." The other simply requested, "Please send me one pound of X-rays and bill as soon as possible."

To clear up such misconceptions, Edison and other scientists set up exhibitions to educate the public first-hand about Roentgen's amazing rays. As it turned out, it was often the scientists who were educated about the public. At one exhibit in London, an attendant reported that two elderly ladies entered the small X-ray room, asked that the door be fastened tightly, and then solemnly requested that he "show them each other's bones, but not below the waistline." As the attendant prepared to comply, a brief argument broke out as "Each wished to view the osseous structures of her friend first." At another point, a young girl asked the attendant if he could take an X-ray of her boyfriend "unbeknown to him, to see if he was quite healthy in his interiors."

Not surprisingly, X-rays brought out the human penchant for foolish hopes and silly deceptions. Columbia College reported that someone had found that projecting X-rays of a bone upon the brain of a dog caused the dog to immediately become hungry. A New York newspaper claimed the College of Physicians and Surgeons had found that X-rays could be used to project anatomic diagrams directly into the brains of medical students, "making a much more enduring impression than ordinary methods of learning anatomic details." And a newspaper in Iowa reported that a Columbia college graduate had successfully used X-rays to transform a 13-cent piece of metal into "$153 of gold."

But to its credit, the public soon recognized that X-rays could be used in equally valuable, but more realistic, ways. A Colorado newspaper reported in late 1896 that X-ray images had been used to settle a malpractice suit against a surgeon who had not properly treated a patient's broken leg. Interestingly, one judge refused to accept the X-ray evidence "because there is no proof that such a thing is possible. It is like offering the photograph of a ghost." But another judge later praised the X-ray evidence and that "modern science has made it possible to look beneath the tissues of the human body."

In the end, perhaps it was a sense of humor that helped society survive that first year following Roentgen's discovery. A political commentary in a 1896 newspaper joked that the Shah had all of his court officials photographed with Roentgen rays and that "In spite of a one hour's exposure, no backbone could be detected in any one of them." In another bit of humor, the *Electrical World* wrote in March 1896 that a woman, apparently fixated on Roman numerals, "recently asked us something about those wonderful 'Ten rays.'" And in August, 1896, the *Electrical Engineer*, mystified by a photographer's ad claiming he could use X-rays to settle divorce cases, wrote, "We presume he uses the X-ray to discover the skeleton which every closet is said to contain."

Finally, a poem that appeared in *Photography* in early 1896 captured the public's nervous amusement with the new rays. Titled "X-actly So!" the poem concluded,

> I'm full of daze
> Shock and amaze;
> For nowadays
> I hear they'll gaze
> Thro' cloak and gown—and even stays,
> These naughty, naughty Roentgen Rays.

Milestone #3 Mapping the unknown country: X-rays revolutionize diagnostic medicine

With all their potential to uncover life-threatening injuries and diseases in virtually any part of the body, it's ironic that the first medical use of X-rays was so remarkably nondramatic: locating a needle. On January 6, just two days after the discovery was announced, a woman came into Queens' Hospital in Birmingham, England, complaining of a sore hand. Fortunately, the necessary equipment was available. An X-ray was made and passed on to a surgeon, who used the image to locate and remove the slender invader. Yet the importance of locating stray needles should not be underestimated, given the apparent frequency of such mishaps. One physicist at Manchester University complained that in early 1896, "My laboratory was inundated by medical men bringing patients who were suspected of having needles in various parts of their bodies. During one week, I had to give the better part of three mornings locating a needle in the foot of a ballet dancer."

But it was not long before physicians began using X-rays for far more serious injuries. In North America, the first use of X-rays for diagnosis and guiding surgery was on February 7, 1896. Several weeks earlier, on Christmas day, a young man named Tolson Cunning had been shot in the leg while playing in a scrimmage. When doctors at Montreal General Hospital couldn't find the bullet, a 45-minute X-ray revealed the flattened intruder lodged between his tibia and fibula. The image not only helped surgeons remove the bullet, but helped Cunning in a law suit he later filed against the shooter. Fortunately or unfortunately, X-rays soon played a starring role in such emergencies. As *The Electrician* wryly observed in early 1896, "So long as individuals of the human race continue to inject bullets into one another, it is well to be provided with easy means for inspecting the position of the injected lead, and to that extent aid the skilled operators whose business and joy it is to extract them."

As X-rays continued to prove their diagnostic value, physicians began demanding that the equipment—often located in a physics laboratory half-way across town—be brought closer to their practices. Thus, as early as April, 1896, the first two X-ray departments in the United States were installed in the New York Post-Graduate Medical School and in the Hahnemann Hospital and Medical College in Chicago. With the opening of the Post-Graduate Medical School facility, *Electrical Engineer* reported that "The utility of taking X-ray pictures in surgery has been demonstrated so often that the hospital authorities have set aside one of the smaller wards for that purpose. They will equip it with Crookes tubes…and all the other paraphernalia of the new art."

X-ray equipment was also soon enlisted for service on the battlefield. In May, 1896, the War Office of the British Government ordered two X-ray machines "to be sent up the Nile to help army surgeons locate bullets in soldiers and in determining the extent of bone fracture." Interestingly, nearly 20 years later, as hospitals were overwhelmed by "terrifying numbers" of wounded soldiers during World War I, Nobel Prize winner

Marie Curie helped expand the use of X-rays and save numerous lives. Curie created what became known as the "petite Curie," a motor vehicle that was equipped with an X-ray machine and powered by the car engine. The vehicle could be driven to the battlefront or short-handed hospitals in and around Paris to help in treating wounded soldiers.

Apart from needles and bullets, X-rays quickly found their way into many other medical applications. One important use was the diagnosis of tuberculosis, a leading cause of death in the late nineteenth and early twentieth centuries. In early 1896, physician Francis Williams—widely considered to be America's "first radiologist"—was diligently at work at Boston City Hospital testing the fluoroscope for its use in diagnosing chest diseases. In April, Williams wrote a letter about his work to a major medical journal, reporting, "One of the most interesting cases was that of a patient suffering from tuberculosis of the right lung…the difference in the amount of rays which passed through the two sides of the chest was very striking…The diseased lung being darker throughout than the normal lung…" In early 1897, after working with other patients and lung diseases, Williams wrote a classic paper in which he concluded, "By X-ray examinations of the chest we gain assistance in recognizing… tuberculosis, pneumonia, infarction, edema, congestion of the lungs in aneurysm, and in new growths…"

The use of X-rays in dentistry was also first reported early in 1896, in this case by William J. Morton (son of William T. G. Morton, who helped discover ether for anesthesia in 1846). In an April meeting of the New York Odontological Society, Morton announced that because the density of teeth is greater than the surrounding bone, "pictures of the living teeth may be taken by the X-ray, even of each wandering fang or root, however deeply imbedded in its sockets." Morton also found that X-rays could be used to locate metal fillings, diseases within the tooth, and even "the lost end of a broken drill." Nevertheless, the regular use of X-rays in dentistry would not come for several decades. With the high voltages, exposed wires, and proximity to the patient's head, the risk of electrical shock—if not electrocution—was literally too close for comfort. Thus, X-rays in modern dentistry did not arrive until 1933, when improved X-ray equipment and dangerous wiring could be enclosed within a smaller unit.

As the diagnostic uses of X-rays expanded, their value was never better appreciated than in emergency situations. In one such case, just months after the discovery of X-rays, a ten-year-old boy accidentally swallowed a nail. When the doctor could not find anything in the boy's throat, he concluded the nail had landed in the boy's stomach and advised the boy to "eat large quantities of mashed potatoes." The boy was fine for a few days, but was then stricken by attacks of coughing. X-ray equipment was called in and, though the first fluoroscopic examination revealed nothing, doctors tried again during one of the boy's coughing attacks. Sure enough, there on the screen, rising up and down a distance of two inches with each cough, was the culprit—not in the boy's digestive tract surrounded by a bolus of mashed potatoes, but lodged in one of his breathing passages. The nail had not been swallowed, but *inhaled*. With the nail located, the physician who reported this case concluded, "Now the surgeon has the last word."

And finally, sometimes X-rays proved to be as valuable in diagnosing conditions of the mind as those of the body. In March, 1896, the *Union Medical* reported that a young woman had requested that her doctor operate on her arm for a pain that she knew was caused by some kind of bone disease. The doctor, who had diagnosed her pain as due to a slight trauma, was proven correct by an X-ray. And thus, "The patient left, entirely cured."

Regardless of how they were put to use, it was soon clear that X-rays would—in fact, *must*—change the practice of medicine forever. On March 6, barely three months after the discovery was announced, Professor Henry Cattell of the University of Pennsylvania wrote in *Science* that, "It is even now questionable whether a surgeon would be morally justified in performing certain operations without first having seen pictured by these rays the field of his work—a map, as it were, of the unknown country he is to explore."

Milestone #4 From hairy birthmarks to deadly cancer: a new form of treatment

"Herr Director, the hair has come out!"

—Leopold Freund, 1896

While these words hardly sound like a promising introduction to a revolutionary medical advance, when they were shouted out by Vienna X-ray specialist Leopold Freund in November, 1896, as he burst into the room of the director of the Royal Research Institute, they marked the first successful use of X-rays as a form of treatment. Yanked by the hand behind Freund was the lucky patient, a small girl who was disfigured by a "tremendous" hairy pigmented birthmark that covered most of her back. Freund had decided to investigate whether X-rays could help her after reading in a newspaper that excessive X-ray exposure could cause hair loss. And indeed, after treating the upper part of the girl's birthmark to X-rays—two hours every day for ten days—the resulting circular bald spot was clear evidence of the therapeutic potential of X-rays.

As Freund and others were beginning to realize, the beneficial effects of X-rays were closely linked their harmful effects. Given the crude equipment and long exposure times being used at the time, the occurrence of harmful effects—including severe burns to the skin and hair loss—hardly seem surprising to us today. Yet it took a milestone insight for early pioneers to investigate such effects as a possible treatment. Interestingly, one of the first people to suggest the treatment potential of X-rays was Joseph Lister, the physician who played a role in the discovery of germ theory. In an address to the Association for the Advancement of Science in September, 1896, Lister noted that the "aggravated sun burning" seen with long exposure to X-rays "suggests the idea that the transmission of the rays through the human body may not be altogether a matter of indifference to internal organs, but may by long continued action produce… injurious irritation *or salutary* stimulation."

In fact, X-rays were soon found to have therapeutic benefits in many skin diseases, including an ability to shrink and dry up the open sores seen in some cancers. What's more, some physicians found that X-rays were particularly good at suppressing pain and inflammation in cancer patients. For example, after using X-rays to treat one patient with oral cancer and another with stomach cancer, French physician Victor Despeignes concluded that "the Roentgen rays have a distinct anesthetic effect, and [provide] a general improvement in the condition of the patient." Similarly, Francis Williams observed that X-rays relieved pain in a breast cancer patient, and that the pain quickly returned when he had to stop treatment for 12 days due to equipment failure.

Although Despeignes also reported that the rays had "little influence" on the growth of cancer, more promising results were seen after X-ray tube technology made a milestone leap in 1913 with the development of the Coolidge tube (discussed later in this chapter). In fact, researchers were eventually surprised to see that higher X-ray energies could kill more cancer cells, while being *less* damaging to normal cells. From this finding came the rationale for the modern treatment of cancer with X-rays: Because cancer cells grow more rapidly than normal cells, they are more susceptible to destruction by X-rays and less capable of regeneration than slower-growing normal cells.

Of course, not everyone limited their efforts to the treatment of serious diseases. In July 1896, the *British Journal of Photography* reported that Frenchman M. Gaudoin, having read that X-rays could cause hair to fall out, made a brief foray into the depilatory business. Gaudoin reportedly hoped to help the "considerable proportion of his country-women endowed with soft silky moustaches, which are by no means appreciated by marriageable young girls and even married ladies." But though customers "flocked" to his business, the treatment failed to work. He was thus forced to "appease their infuriated graces" by returning their money and then "hurriedly retired from the business."

Milestone #5

A darker side is exposed: the deadly dangers of X-rays

One summer day in 1896, William Levy, intrigued by reports of the miracle new rays, decided it was high time to look into getting that pesky bullet removed from his brain. Shot just above his left ear ten years earlier by an escaping bank defaulter, Levy survived the attack and now approached a professor at the University of Minnesota to see if X-rays could help doctors locate and remove the bullet. Levy was duly warned that the long exposures needed to penetrate his skull might cause some hair loss. And so, on July 8—in a marathon *14-hour session*—he sat while X-ray exposures were taken at various points around his head, including one *inside* his mouth. Levy suffered no pain, but within days, his skin turned an angry red and began to blister, his lips became swollen, cracked, and bleeding, his mouth was so burned he could only ingest liquids, and his right ear swelled grotesquely to twice its normal size. And, oh yes, the hair on the right side of his head fell out. The good news was that not only did two X-ray images reveal the location of the bullet, but within four months Levy had recovered sufficiently to ask the professor to take *more* X-rays to help doctors determine the feasibility of an operation.

Throughout 1896, reports of side effects like those experienced by Levy provided growing evidence that Roentgen's invisible rays were not simply passing harmlessly through the body. Some scientists, doubting that X-rays were to blame, suggested instead that the burns and hair loss were caused by the electrical discharges needed to produce the rays. Thus, it was proposed that the injuries might be avoided if X-rays were produced instead by "static" machines. But it did not take long for scientists—wielding their own swollen and burned fingers as proof—to show that the X-rays emitted by these machines were just as harmful. And so within a year of their discovery, it was increasingly clear that X-rays could cause short-term damage to tissues. What nobody yet suspected, however, was that the rays might cause *long-term* effects.

That direct exposure to X-rays could damage the body is hardly surprising given that exposure times in the early years were often an hour or longer. And of course patients were not the only ones at risk. One of the tragedies of early X-ray research is that it was often the scientists and clinicians who—having exposed their own hands and fingers to the rays day after day—suffered first and worst. One famous case was that of Clarence Dally, who assisted Thomas Edison in his early work in X-rays and often held objects beneath the rays with no protection. Dally eventually developed severe burns to his face, hands, and arms. In 1904, despite amputation of both his arms in an attempt to stop recurrent cancer, Dally died. While this tragic event helped alert the world to the dangers of X-rays, it also prompted Edison to abandon his X-ray research, despite his pioneering work in developing the fluoroscope and other achievements.

Interestingly, some early pioneers, thanks to a combination of intuition and luck, managed to escape harm. Roentgen, for example, conducted many of his experiments in a large zinc box, which provided the necessary shielding. And Francis Williams protected himself from the very start of his work because, as he later explained, "I thought that rays having such power of penetrating matter must have some effect upon the system, and therefore I protected myself."

Unfortunately, the early years of unshielded X-ray use eventually took their toll on many early pioneers. In 1921, following the deaths of two famous radiologists in Europe, the *New York Times* published an article about the dangers of unprotected exposure to X-rays, listing a number of radiographers and technicians who had died between 1915 and 1920. Many of them, like Dally, endured multiple operations and amputations in a futile attempt to stop the spread of cancer. And some were heroic in facing the inevitable. After suffering facial burns and the amputation of his fingers, Dr. Maxime Menard, chief of the "electrotherapeutic" department at a hospital in Paris, reportedly said, "If the X-rays get me, at least I shall know that with them I have saved others."

Eventually, a new understanding of X-rays and their biological effects helped clarify the risks. As we now know, X-rays are a form of light (electromagnetic radiation) so intensely energetic that they can strip electrons from atoms and thereby alter cellular functions at a molecular level. Thus, when X-rays pass through the body, they can have one of two major effects on cells, either killing them or damaging them. When cells are killed, short-term adverse effects such as burns and hair loss may occur. But if the X-rays "merely" damage DNA *without* killing the cell, the cell can continue to divide and pass the mutated DNA on to daughter cells. Years or decades later, these mutations can lead to the development of cancer.

Fortunately, by 1910, the hidden dangers of X-rays had been exposed, and protective goggles and shielding were used with growing frequency by scientists and clinicians. Having passed this dark milestone, X-rays could now move on to an even brighter and safer future in medicine.

Milestone #6

A leap into the modern age: Coolidge's hot tube

From the day Roentgen first announced his discovery, scientists following in his footsteps began tinkering with various components of the equipment in attempts to make X-ray images sharper, shorten exposure times, and achieve better penetration of the body. It was one thing to create images of bones in the hands, which are relatively thin, flat, and easy to hold still for long exposures; capturing images of organs deep in the chest and abdomen was far more challenging. While a succession of technical improvements during the first decade or so allowed radiographers to create X-ray images of numerous body organs, image quality and exposure times remained a key limitation. And these were largely due to the design of the X-ray tube itself.

The basic problem with early tubes such as the Crookes tube was they were not true vacuum tubes: The tubes always contained some residual gas molecules. This was both good and bad. On the one hand, gas molecules were *needed* to create X-rays, given that it was their collision against the cathode that created cathode rays, which in turn created X-rays. On the other hand, the residual gas molecules were a problem because with repeated use, they altered the composition of the glass tube itself and disrupted its ability to produce X-rays. While the altered tubes produced more penetrating X-rays, the *intensity* was decreased, resulting in poorer image quality. The end result was that, over time, X-ray tubes became erratic—so much so that Roentgen once wrote in a letter, "I do not want to get involved in anything that has to do with the properties of the tubes, for these things are even more capricious and unpredictable than women."

While many clever designs were implemented to compensate for the technical limitations of early X-ray tubes, the true milestone—what some experts call "the single most important event in the progress of radiology"—did not occur until nearly 20 years later. In 1913, William Coolidge, working in the General Electric Research Laboratory, developed the first so-called "hot" X-ray tube, subsequently called the Coolidge tube. Based on his earlier research, Coolidge had figured out how to make the cathode out of the metal tungsten, which has the highest melting point of all metals. With a cathode made primarily of tungsten, cathode rays could be generated by running an electric current through the cathode and heating it; the more the cathode was heated, the more cathode rays it emitted. Thus, with cathode rays generated by heat rather than gas molecule collisions, the Coolidge tube could operate in a perfect vacuum.

Thanks to these and other design changes, the Coolidge tube was not only more stable—producing consistent, reliable exposures—but operators could also now independently control X-ray intensity *and* penetration. X-ray intensity was controlled by changing the temperature of the cathode, while penetration was controlled by changing the tube voltage. Finally, by operating in a true vacuum, Coolidge tubes were less finicky and could function almost indefinitely, unless broken or badly abused.

By the mid-1920s, the Coolidge tube had essentially replaced the old gas-filled tubes. In addition, Coolidge later designed other innovations so that higher voltages could be used to produce higher frequency X-rays. This led to the development of so-called "deep therapy," in which X-rays are used to treat deeper tissues without excessively damaging outer layers of skin. Thanks to Coolidge's milestone redesign of the X-ray tube, the use of X-rays in medicine—for both diagnostic and therapeutic applications—expanded widely throughout the world from the 1920s and onward. Today, Coolidge's "hot" tube design is still the basis for all modern X-ray tubes.

Milestone #7

A final secret revealed: the true nature of X-rays

If you were a scientist or layman in 1896 and fascinated by the discovery of X-rays, you probably would have been equally intrigued, if not amused, by some of the theories attempting to explain what they were. For example, physicist Albert A. Michelson curiously suggested that they were "electromagnetic whirlpools swirling through the ether." And there was Thomas Edison's proposal, eventually discredited as "nonsense," that X-rays were "high-pitched sound waves." Other theories included the view that—despite evidence to the contrary—X-rays were actually cathode rays.

Interestingly, Roentgen was closer to the mark in his landmark 1895 paper when he observed that X-rays were similar to light because, for example, they created images on photographic film. Yet he also observed that X-rays were different than light because they could not be diffracted by a prism or "bent" by magnets or other substances. With these and other contradictory observations, the mysterious nature of X-rays entered the greater debate among physicists at the time as to whether light was made up of particles or waves. But before long, increasing evidence suggested that X-rays were indeed a form of light—that is, a form electromagnetic radiation that traveled through space in waves. Roentgen and others had been initially misled because the wavelengths of X-rays are so incredibly short—in fact, *10,000 times* shorter than visible light.

The final proof came on April 23, 1912, when physicist Max von Laue performed a milestone experiment. Von Laue had been contemplating how to show that X-rays were truly electromagnetic waves and—in what might seem to be an unrelated problem—whether the atoms in a crystal are arranged in a regular lattice-like structure. In a brilliant insight, von Laue addressed both questions with a single experiment. He sent a beam of X-rays through a crystal of copper sulphate, theorizing that if the atoms were indeed structured as a lattice—and if X-rays were indeed composed of waves—the spacing between the atoms might be sufficiently small to diffract the tiny X-ray waves. Von Laue's experiment confirmed both theories. Based on the distinctive "interference" pattern the X-rays made when they emerged from the crystal and struck a photographic plate, von Laue was able to deduce that the atoms in a crystal are indeed arranged in a lattice *and* that X-rays travel in waves and are therefore a form of light. For his milestone discovery, von Laue received the 1914 Nobel Prize in Physics.

The twentieth century and beyond: the milestones keep on coming

While the milestones discussed here represent the most important advances in the discovery and application of X-rays to medicine, in recent years newer milestones have continued the revolution. Some of these milestones, such as the development of contrast agents, are so broad that they apply to many areas of diagnostic radiology. Others are specific to a given body region but still have profoundly impacted medicine and health. One example is mammography, the use of low-dose X-rays to detect and diagnose breast cancer. Although X-rays were first used to examine breast disease in 1913 by German surgeon Albert Salomon, the initial techniques were crude and unreliable. In 1930, radiologist Stafford Warren was one of the first investigators to provide reliable data on the clinical use of breast X-rays. But it was not until 1960 that Robert Egan, a radiologist at the University of Texas, published a landmark study in which he described mammography techniques that could attain 97% to 99% accuracy in detecting breast cancer. Egan's results proved the validity of mammography and led to its widespread use in breast cancer screening. By 2005, mammography accounted for 18.3 million office visits in the United States, or about 30% of all X-ray exams.

But perhaps the most astonishing recent milestone was the development of an entirely new way of using X-rays to reveal the inner world of the body. Until the 1970s, all X-ray images had one major limitation: They were flat and two-dimensional. Lacking depth, X-ray images of internal organs are often obscured by overlapping organs and tissues that cause unwanted shadows and reduced contrast. This is why physicians, in an effort to gain additional perspective, often order two X-ray images (one from the front and one from the side). But in 1971, British engineer Godfrey Hounsfield overcame this limitation with the development of computed tomography (CT), in which X-rays are used to take a series of cross-sectional images, or "slices," of the body area being examined. (*Tomos* is a Greek word that means to cut or section.) With CT, instead of sending a single beam through the body to create a single image, X-rays are sent through the patient multiple times from multiple angles around the body and collected by detectors that convert them into electrical signals. These signals are then sent to a computer, which reconstructs the data into detailed cross-sectional "slices" that can be assembled into three-dimensional images. Because the image data doesn't overlap while the image is being constructed and because CT detectors are more sensitive than film, CT can show much finer variations of tissue density than conventional X-rays.

The development of CT was aided by two key developments in the 1960s and 1970s. One was the advent of powerful minicomputers, which were needed to process the enormous amount of data captured by X-ray detectors and reconstruct it into images. The second was the work of Alan Cormack, who created a mathematical model for measuring different tissue densities in the body and predicting how this information could be used to create cross-sectional X-ray images. For their work in developing CT, Hounsfield and Cormack were awarded the 1979 Nobel Prize in Physiology or Medicine.

In its early uses, CT produced the first clear images of the brain's gray and white matter and thus had a major impact on the diagnosis of neural diseases. Since then, numerous advances have led to faster scanning, thinner slices, and the ability to scan larger body areas. Today, CT scanners can produce exquisite, 3D images in virtually every part of the body. One recent application, for example, is virtual colonoscopy, in which CT produces images of the interior of the large intestine. Less invasive than the conventional method of threading a long, flexible optical tube through the colon, virtual colonoscopy is becoming an increasingly important tool in screening for colon cancer.

A remarkable range of uses—but always medicine's trustworthy guide

Apart from their role in medicine, X-rays have had a major impact in numerous other areas of science and society. Within years of their discovery, X-rays were put to use in many areas of industry, including detecting flaws in iron castings and guns, examining submarine telegraph cable insulation, inspecting the structure of airplanes, and even examining live oysters for pearls. X-rays have also found important applications in basic biology (revealing the structure of proteins and DNA), the fine arts (detecting fraudulent imitations of paintings), archeology (assessing objects and human remains at archeological sites), and security (inspection of baggage, packages, and mail).

But for their sheer impact on saving or improving human lives, X-rays have made their greatest impact in the field of medicine. According to the Centers for Disease Control and Prevention (CDC), X-rays are still one of the most common medical tests. For example, in 2005 X-rays were ordered in 56.1 million office visits, making them nearly twice as common as ultrasound, MRI, and PET imaging tests. In terms of frequency of diagnostic tests given in an office visit, X-rays rank behind only three major blood tests (CBC, cholesterol, and glucose) and urinalysis.

Of course, X-rays have important limitations that cannot be overlooked. Today, for example, they are often used with or replaced by other imaging technologies, such as MRI, ultrasound, and PET, to provide anatomic and physiologic insights that X-rays alone could never achieve. In addition, the cumulative effects of X-rays continue to be a concern and play a role in determining the viability of some new applications. For example, CT angiograms are a promising new tool for non-invasively examining the coronary arteries and assessing heart disease risk. However, CT angiograms can expose patients to the equivalent of at least several *hundred* standard X-rays, thereby posing a small but real cancer risk. Thus, as with all evolving technologies, even the most exciting milestones must be continually evaluated for their balance of risks and benefits.

Nevertheless, it would be a shame if we ever lost that sense of awe and appreciation that shook the world when X-rays were first discovered, when those tiny, invisible waves of light first began opening unimagined new views into the human body. Nor should we forget that sometimes the images they unveil—whether bullets, bones, needles, or cancer—have little to do with hidden mysteries or secrets.

Take the case of the 39-year-old construction worker who was injured in 2004 in a grisly nailgun accident. There was no mystery when the nailgun suddenly fired six 3-inch nails into his face, spinal column, and skull, sending him to a Los Angeles hospital fearing for his life. Or the 59-year-old German woman who, when she was just four years old, fell down while carrying a 3-inch pencil. There was no secret when the pencil pierced her cheek and disappeared into her head, causing a lifetime of headaches, nosebleeds, and loss of smell. But in both cases, X-rays helped doctors locate the invaders and perform the surgery needed to successfully remove them.

In such cases, X-rays are more like a trustworthy guide, the same reliable friend that first helped doctors remove a needle from a woman's hand, just two days after the announcement of their discovery. More than a century later, in both diagnosis and treatment, X-rays continue to provide the roadmaps and tools doctors need to save or improve the lives of millions.

The Scratch that Saved a Million Lives: THE DISCOVERY OF VACCINES

6

Clara and Edgar, Part I

Riding on the wave of an explosive sneeze, the microscopic enemy blasts out at 100 miles per hour, a cloud of 40,000 aerosolized droplets that instantly fills the room. Clinging to ocean-sized droplets, the invisibly small microbes drift about for several minutes, patiently awaiting their next victim. The wait is not long. Gently wiping the nose of her dying four-year-old child, Clara complies by the simple act of breathing.

The enemy makes landfall in Clara's nose and throat and within hours has mobilized into nearby lymph nodes. Entering cells, it converts them into reproductive slaves, forcing them to churn out the enemy's own offspring. In just half a day, each cell begins releasing a new generation, dozens of new invaders to join the expanding army, infecting more and more cells. Several days later, the enemy enters Clara's bloodstream.

But as the deadly assault continues, Clara's own protective army fails to mount a response. The enemy slips quietly by, unrecognized and undeterred…

* * *

The "enemy" is *variola* virus—smallpox—a fuzzy, brick-shaped microbe so tiny that a bacterium would tower over it like a small house, a red blood cell would dwarf it like a football stadium. Like other viruses, it is so genetically primitive that it exists in the creepy netherworld between the living and the dead. For tens of thousands of years, its ancestors lived in Africa, content to infect only rodents. But about 16,000 years ago, something within its sparse 200 genes mutated and gave birth to a new form—a virus that could infect *only* humans. From that time onward, the new strain ungraciously thanked its human hosts by killing 30% of those it inhabited.

Over thousands of years, the virus joined its human hosts in their migration out of Africa, into Asia, and eventually Europe. With each sick person capable of infecting five or six others, it traveled easily from culture to culture, triggering waves of epidemics. The first evidence of its presence in humans is seen in skin rashes on Egyptian mummies dating from 1580 BC. The first recorded smallpox epidemic occurred 200 years later during the Egyptian-Hittite war. By 1122 BC, smallpox-like diseases were being reported in ancient China and India…

Clara and Edgar: Part II

The enemy continues to multiply relentlessly throughout Clara's body, but only now—two weeks later—does she experience her first symptoms. It begins with a high fever, chills, and exhaustion. Then severe headaches, backache, and nausea. Mercifully, the symptoms abate within days—just in time for the real trouble to begin. Mounting a whole new campaign of terror, the virus begins invading small blood vessels in her skin.

The infamous rash—the "Speckled Monster"—first emerges as small red spots on the tongue and in the mouth. Soon, it appears on her face, and within 24 hours it has spread over her entire body. For the next few weeks, the rash follows a hideous and predictable progression: Flat red spots rise into bumps. The bumps fill with a thick milky fluid that develop belly button-like depressions and then evolve into rounded pustules so packed with fluid that they feel like hundreds of beads embedded in the skin. As pustules erupt over the entire body and emit a repugnant smell, the effect is grotesque, as if evil itself were bubbling up from inside. Finally, the pustules dry into a crust and form a scab. When the scabs fall off, they leave a disfiguring landscape, a face pitted by scars…

But all of this assumes that Clara is still alive. In one-third of cases, often when the pustules are so widespread that they touch one another, patients die from an immune system so overwhelmed that it destroys the very tissues it is trying to save. The virus also attacks other parts of the body, leaving many survivors blind and with limb deformities. In the meantime, anyone close to the patient during the contagious rash phase may already be nursing the next generation.

* * *

One of the first and most devastating smallpox epidemics to be recorded, the Antonine plague, began in 165 AD and lasted until 180 AD. Killing from three to seven million people, some theorize that it contributed to the fall of the Roman Empire. As the centuries rolled on, the virus continued its deadly global march by joining the Crusades and Arab expansion. By the 1500s, it threatened—and actually eliminated—entire civilizations. Brought to the New World by Spanish and Portuguese conquistadors, smallpox killed 3.5 million Aztec Indians and caused the downfall of both the Aztec and Incan empires. In the eighteenth century, smallpox was endemic or epidemic in most major European cities, killing

400,000 people a year, including five reigning European monarchs, and causing up to one-third of all blindness.

Clara and Edgar: Conclusion

Having buried their child a few days earlier, Edgar enters the room to care for his dying wife, Clara. He finds it unbearable to watch her agonizing struggle, recalling how he once suffered from the same illness when he was a child. In the final hours before Clara's death—just two weeks after her symptoms began—there is an explosive sneeze and the simple act of breathing in the air. The enemy makes landfall in Edgar's nose begins its next invasion.

But this time the virus is not so lucky. Inside Edgar's body, it is immediately recognized by cells that remember his childhood encounter from long ago. The cells spring to life, multiply, and begin producing a deadly weapon: antibodies. Specialized proteins designed to target and attack this exact invader, the antibodies begin their work. They block the virus from latching onto cells; they stop it from entering cells; they prevent it from replicating in cells, and—for any that manage to survive—they neutralize and help destroy it. In the weeks that follow, Edgar does not experience a single symptom.

* * *

One of the first major clues that smallpox could be defeated was recorded in 910 AD by the Persian physician Rhazes (Al-Rhazi). Considered the greatest physician of the Islamic world, Rhazes not only wrote the first known medical account of smallpox, but noted a curious—and critical—clue: People who survived smallpox were somehow *protected* against subsequent attacks.

About the same time, writings began to appear in China providing a second key clue: People could actually protect themselves from the disease by taking scabs from a victim, crushing them into a powder, and swallowing or scratching it into the skin. But though this unsavory-sounding practice—called variolation—seemed to work and eventually was also practiced in Asia and India, it was not widely adopted, perhaps because of one unfortunate side effect: The risk of accidentally contracting full-blown smallpox and dying.

And so the deadly epidemics continued over the centuries, spreading and periodically erupting around the globe. For 16,000 years, tragic stories like that of Clara and Edgar recurred in countless variations as the smallpox virus conducted its relentless death march through human civilization. Until finally in the late eighteenth century, a country doctor in Gloucestershire, England, performed a curious experiment that would change the world...

May 14, 1796: An historic turn of events

James Phipps, a healthy young eight-year-old boy, is brought into a room where a doctor suddenly seizes his bare arm and cuts two shallow incisions into his skin. A cloud of tiny particles—taken from a sore on the hand of a dairymaid infected by a disease called cowpox—instantly fills the shallow wound.

Making landfall near the base of James' epidermis, the microscopic enemy—the cowpox virus—enters nearby cells and begins replicating. But despite its distant relation to smallpox, this virus poses little danger. Within days, specialized cells in James' body begin producing antibodies that target and attack the invader. The cowpox virus is quickly defeated, and James experiences only mild symptoms. But as the evidence will later show, James is not only protected from future attacks by cowpox: Because of the virus's similarity to its deadly cousin, he is now also immune to smallpox.

* * *

Although it would be nearly another 100 years before scientists had even a rudimentary understanding of *why* it worked, when Edward Jenner inoculated James Phipps in May, 1796 with infectious cowpox viruses from a lesion on the hand of a dairymaid, he capitalized on clues that had been accumulating for more than 1,000 years. And in so doing, he laid the scientific foundation for one of the greatest breakthroughs in medicine: vaccines.

Vaccines' clever secret: not fighting, but *teaching* the body to fight disease

It is fortunate for the human race that the world's first vaccine was so effective against what many consider to be the world's worst disease. Few people today remember the threat smallpox once posed to human civilization, but even as late as the 1950s—150 years *after* the discovery of an effective vaccine—smallpox continued to infect 50 million people yearly, killing *two million* of them. As the World Health Organization has noted, no other disease of the past or present has approached smallpox in the devastation it has caused the world population.

Since Jenner's historic milestone 200 years ago, advances in vaccines have followed a long and remarkable road, reflecting the complexity of disease and the intricacies of the human body. Today, vaccines remain one of medicine's most remarkable approaches to fighting disease for two reasons. First, unlike most treatments, vaccines do not directly attack disease. Rather, they *teach the body* how to fight a disease by training it to produce its own weaponry—antibodies. Second, in a delicious twist of biological irony, every vaccine is made from the very disease it is designed to fight—typically, a weakened or killed form of the bacteria or virus in question.

While the journey to understanding and creating vaccines was initially slow, soon a dizzying series of milestones would create a growing arsenal of vaccines. Today, vaccines have enabled us to control ten major diseases—smallpox, diphtheria, tetanus, yellow fever, pertussis, *Haemophilus influenzae* type b, polio, measles, mumps, and rubella (German measles).

But while Edward Jenner is properly credited for his role in the discovery of vaccines, often overlooked is the fact that the first true milestone occurred decades earlier in southern England, when a farmer named Benjamin Jesty took a daring risk to save his family from a local outbreak of smallpox. Jesty led his wife and two young children on a two-mile hike through a patchwork of hedgerows near the wooded slopes of Melbury Bubb and the River Wiggle. And there, in the cow pastures of Farmer Elford, he gathered his family, knelt down by a sick cow, and pulled out a sharp stocking needle…

Milestone #1 What the milkmaid knew: a daring experiment in a cow pasture

In some ways, it's surprising no one had thought of it before. On the one hand, by the mid-1700s it was common knowledge among country folk that milkmaids who caught relatively mild cowpox were subsequently immune to deadly smallpox. Cowpox was an annoyance with which many farmers were familiar: Sporadically erupting on farms, it caused small pustules to form on the udders of cows and decrease their milk production. This did not please the farmers, nor did the fact if one of their milkmaids contracted the disease through an open cut, she would soon develop similar pustules on her skin, along with fever, headaches, and other symptoms that forced her to stop working for a few days. Fortunately, the dairymaids quickly recovered, and when they did, they were now immune to not only cowpox but—if folklore was to be believed—smallpox.

On the other hand, variolation—the risky practice of inoculating people with a small amount of live smallpox to protect them against it—had been introduced to England in 1721 and by the mid-1700s was well-known and practiced by many physicians. Yet there remained a critical gap: Few people made the link between what dairymaids knew about cowpox and what doctors knew about smallpox variolation…until Benjamin Jesty's historic family field trip to a nearby cow pasture.

Benjamin Jesty was a prosperous farmer who, despite a lack medical training, had a reputation for intelligence and a penchant for innovation. Thus, in 1774, when a smallpox epidemic broke out in Jesty's community in Dorset county, a fear for his family's health set him to thinking. Although not everyone believed—or was even aware—that cowpox might protect a person from smallpox, Jesty had heard the rumors. In fact, years before, two servants, who both had previously contracted cowpox, had told him that they had later survived an outbreak of smallpox, despite caring for two boys with the highly contagious disease. Jesty filed this information away, just as he had filed away something else he had apparently learned from local doctors: the technique of variolation.

And so in the spring of 1774, putting those two facts together, 37-year-old Jesty took a leap of faith no one else had made. With a local smallpox outbreak under way, he led his family on a two-mile walk through the patchwork hedgerows and wooded slopes of Melbury Bubb, entered Farmer Elford's pasture, and found a cow whose udders were marked by the distinctive sores of cowpox. Jesty then pulled out one of his wife's stocking needles, dipped its slim point into an open lesion, and did something most people at the time would have considered ill-advised, if not immoral. He inoculated his entire family with the infectious cowpox material: immediately below the elbow of his wife Elizabeth (to avoid her dress sleeve) and above the elbows of his sons, Robert, 3, and Benjamin, 2. Jesty did not inoculate himself since he'd already had cowpox as a youth.

The experiment was nearly a disastrous failure. Within days, Elizabeth's arm became inflamed, she developed a serious fever, and might have died had she not received care from a doctor. But happily, Elizabeth recovered, and the experiment proved a success. Jesty's wife and two sons remained free of smallpox for their rest of their lives, despite later exposure to epidemics. What's more, both of his sons' immunity was confirmed when they were later variolated and had no reaction. (A lack of a reaction to variolation is evidence that a person is immune to smallpox.)

Unfortunately, when news of Jesty's experiment got out, it reportedly caused "no small alarm" in the neighborhood, particularly among those who viewed the mixing of substances between humans and animals to be an "abomination" against God. As news spread through the community, whenever Jesty attended local markets he was scorned, ridiculed, and even pelted with mud and stones.

Sadly, despite his successful gamble, Jesty never inoculated another person, and there is no written evidence that Edward Jenner even knew of Jesty's experiment. Nevertheless, Jesty eventually received credit for his milestone, while Jenner would take the discovery to a whole new level—one that would eventually impact the world.

Milestone #2 From folklore to modern medicine: Jenner discovers the science of vaccination

What motivates a man to discover one of the top ten breakthroughs in the history of medicine? In the case of Edward Jenner, it was not simply a desire to conquer the deadliest disease in human history. Rather, it was the desire to spare others from something that had nearly killed him when he was just eight years old: a poorly conceived—if not downright bizarre—attempt to *prevent* the disease.

The irony is that when Jenner was variolated in 1757, the procedure had been practiced in England for 35 years and was considered reasonably safe and well accepted. While variolation clearly had its risks—about 1 in 50 people contracted full-blown smallpox from the procedure and died—it was still preferable to the 1 in 3 risk of death faced by those who contracted smallpox naturally. Nevertheless, in a misguided attempt to improve variolation, some physicians had begun devising "preparations," in which, prior to being variolated, patients were subject to weeks of purging, enemas, bleeding, fasting, and dieting. The ordeal was so extreme that the preparation itself sometimes proved fatal. Narrowly missing this fate as a child, Jenner later recalled of his six-week regimen, "There was taking of blood until the blood was thin, purging until the body was wasted to a skeleton, and starving on a vegetable diet to keep it so."

But Jenner's pain was humanity's gain. Thanks to his frightful experience, he developed a lifelong aversion to variolation and a powerful motivation to find a better way to prevent smallpox. Like Benjamin Jesty, the pieces to the puzzle came to Jenner gradually over the course of many years. Born in Gloucestershire in 1749, one of Jenner's first major clues came to him when he was just 13 years old. Working as an apprentice to a surgeon at the time, Jenner was intrigued when he overheard a dairymaid boasting, "I shall never have an ugly pockmarked face." She was referring, of course, to the facial scarring often seen in those lucky enough to survive smallpox. The reason for her confidence? "I shall never have smallpox," she explained, "for I have had cowpox."

The dairymaid's confidence in local folklore made a strong impression on the young Jenner. From that time forward, he had a persistent curiosity about the link between cowpox and smallpox. Unfortunately, this persistence was not shared by peers, as seen early in his career, when Jenner raised the topic one too many times at an informal medical society. According to his friend and biographer Dr. John Baron, "It became so distasteful to his companions [that] they threatened to expel him if he continued to harass them with so unprofitable a subject."

In 1772, after completing his medical training, 23-year-old Jenner set up a medical practice in Berkeley, Gloucestershire. Around 1780, still intrigued by the connection between cowpox and smallpox, he began collecting case reports of people who had been infected with cowpox and were later found to be immune to smallpox (as seen by their lack of reaction to variolation). In 1788, Jenner made sketches of cowpox lesions he had seen on the hands of infected milkmaids, took them to London to show to several doctors, and discussed his idea that cowpox could protect against smallpox. Most were unimpressed. Similarly, when Jenner later asked some medical colleagues for help in investigating the link, they insisted the idea was ridiculous and merely an old wives' tale.

But Jenner remained undiscouraged and continued his investigations until he could no longer avoid his milestone destiny: On May 14, 1796, taking the matter into his own hands, Jenner performed his first vaccination on eight-year-old James Phipps, the son of a laborer who occasionally worked for Jenner. Jenner inoculated the boy with infectious cowpox "matter" taken from the hand of a milkmaid named Sarah Nelmes, who had picked up *her* infection from a cow named Blossom. Like Benjamin Jesty, who had achieved the same milestone 22 years earlier, Jenner's experiment proved to be a success: When Phipps was variolated six weeks later, and again a few months after that, the lack of reaction showed that he was indeed immune to smallpox. In fact, Phipps went on to live a long life free of smallpox and even had himself variolated some 20 times to prove his immunity.

Yet despite Jenner's victory, news of his success was no more welcome than that of Jesty's 20 years earlier. When in 1796 he submitted a paper to the Royal Society describing the Phipps experiment and 13 case histories of people who became immune to smallpox after immunization with cowpox, the paper was promptly rejected due to a lack of sufficient data. What's more, Jenner's experiment was deemed "at variance with established knowledge" and "incredible," and Jenner was warned that he "had better not promulgate such a wild idea if he valued his reputation."

Jenner could do nothing about the "wild" or "incredible" nature of his idea, but he *could* gather more data. Unfortunately, he had to wait more than a year for the next outbreak of cowpox, but when it finally occurred in the spring of 1798, Jenner inoculated two more children. He then inoculated several more children from lesions on one of the first two children—the so called "arm-to-arm" method. When variolation later showed that all of the children were immune to smallpox, Jenner knew his case was proven. But rather than approach the Royal Society again, he self-published his findings in a now-classic 64-page paper, *An Inquiry into the Causes and Effects of Variolae Vaccinae, or Cowpox*.

Making the leap: from publication to public acceptance

Jenner made many important claims in this paper, including: inoculation with cowpox protected against smallpox; protection could be propagated from person to person by "arm–to-arm" inoculation; and that unlike smallpox, cowpox was not fatal and only produced local, noninfectious lesions. The paper also included Jenner's first use of the term vaccinae (from the Latin *vaca*, meaning cow), from which "vaccine" and "vaccination" would be derived.

Yet even with these new findings, Jenner continued to face doubts and derision from his peers. Opposition ranged along many fronts, with some physicians disputing that cowpox was a mild disease, others claiming that vaccination didn't work when they tried to repeat Jenner's experiment, and still others opposing vaccination on religious or moral grounds. Perhaps the most bizarre objection came from those who claimed that their attempts at vaccination resulted in patients developing "bovine" characteristics—a notion that led to one cartoon showing vaccinated babies with "cow horns" sprouting from their heads.

Eventually, however, as more credible doctors tried the technique, more positive reports began to emerge. The vaccine seemed to work after all, although debates continued over its effectiveness and safety. In the meantime, Jenner continued his work and published more papers that clarified or revised his views based on new evidence. While Jenner was not always right—he incorrectly believed that vaccination provided life-long protection—the practice of vaccination began to spread surprisingly fast. Within a few years, vaccinations were being administered not only in England, but throughout Europe and soon in other areas of the world. In America, the first vaccinations were given on July 8, 1800, when Benjamin Waterhouse, a professor at Harvard Medical School, inoculated his 5-year-old son, two other children, and several servants. Waterhouse later sent the vaccine to President Thomas Jefferson for distribution to southern states, and Jefferson soon had his entire family and neighbors (some 200 people) vaccinated.

By 1801, Jenner had no doubts about the success of vaccination, as seen when he wrote, "The numbers who have partaken of its benefits throughout Europe and other parts or the globe are incalculable, and it now becomes too manifest to admit of controversy that the annihilation of the Small Pox, the most dreadful scourge of the human species, must be the final result of this practice."

Although no one in Jenner's time remotely understood how vaccines worked or what caused smallpox, and though not technically the "first" person to vaccinate a person against smallpox, today historians give Jenner primary credit for this milestone because he was the first to scientifically demonstrate that vaccines can work. Equally important, he gave the world its first reasonably safe way to stop the deadliest disease in human history.

* * *

And yet...despite Jenner's success, it was soon clear that his vaccine had some serious shortcomings. For one thing, the immunity was not lifelong, and no one was exactly sure why. Some scientists speculated that the loss of potency might be due to a concept called "passages," the progressive weakening that occurred as the vaccine was continually transferred through "arm-to-arm" inoculation. In other words, perhaps the "agent" responsible for conferring immunity somehow lost more and more of its disease-fighting ability each time it was transferred from person to person.

At the same time, Jenner's vaccine raised other vexing questions. For example, why couldn't his approach—using a harmless disease to make vaccines against a related dangerous one—be used against *all diseases*? The answer, as we now know, is that Jenner's vaccine was a stroke of enormous good fortune. The fact that the smallpox virus happened to have a harmless but closely related cowpox cousin was a quirk of nature, so rare that it is not seen in any other human infections. Indeed, lacking any other ways to make vaccines, the story of vaccines would have been a very short one indeed.

Perhaps that's why vaccine development *did* hit a dead end for the next 80 years. Until finally one scientist—already a key player in the discovery of "germ" theory—made a milestone leap by going on a long vacation...

Milestone #3 A long holiday and a neglected experiment lead to a new concept in vaccines

By the 1870s, Louis Pasteur had already achieved his lion's share of medical milestones. Over the previous three decades, he had contributed significantly to the discovery of germ theory through his work in fermentation, pasteurization, saving the silkworm industry, and disposing of the theory of spontaneous generation. But in the late 1870s, Pasteur was poised yet again to make a milestone discovery, this time after receiving a rather inauspicious gift: the head of a chicken.

It was no threat or cruel joke. The chicken had died from chicken cholera—a serious and rampant disease that at the time was killing as many as 90 out of 100 chickens—and the veterinarian who sent the specimen to Pasteur for investigation believed the disease was caused by a specific microbe. Pasteur soon verified this theory: When microbes from the chicken head were grown in a culture and then injected into healthy chickens, the injected chickens soon died of chicken cholera. Though this discovery helped support growing evidence for germ theory, Pasteur's culture of bacteria soon played an even more profound role—thanks to a combination of neglect and serendipity.

In the summer of 1879, Pasteur went on a long holiday, forgetting about the chicken cholera culture he had created, and leaving it exposed to the air. When he returned from his vacation and injected it into some chickens, he found the culture was not so deadly anymore: When chickens were inoculated with the weakened, or attenuated, bacteria, they got sick but did *not* die. But Pasteur's next discovery was even more significant. If the chickens were allowed to recover and then were injected with *deadly* chicken cholera bacteria, they were now *fully resistant* to the disease. Pasteur immediately realized that he'd discovered a new way to make vaccines: Inoculation with a weakened form of a microbe somehow enabled the body to fight off its *deadly* form. Discussing his discovery in an 1881 article in *The British Medical Journal*, Pasteur wrote, "We touch the principle of vaccination…When the fowls have been rendered sufficiently ill by the attenuated virus… they will, when inoculated with virulent virus, suffer no evil effects… chicken cholera cannot touch them…"

Inspired by his milestone discovery, Pasteur began investigating how this new approach could be used to make vaccines against other diseases. His next success was with anthrax, a disease that was disrupting the agriculture industry by killing from 10% to 20% of sheep. Earlier, Robert Koch had already shown that anthrax was caused by bacteria. Pasteur now began investigating whether anthrax bacteria could be sufficiently weakened to make them harmless, yet still able to stimulate protection in the body if inoculated as a vaccine. He eventually succeeded by growing the bacteria at elevated temperatures. And when faced by some peers who doubted his findings, Pasteur soon found an opportunity to prove himself with a dramatic public experiment. On May 5, 1881, Pasteur inoculated 24 sheep with his new attenuated anthrax vaccine. Nearly two weeks later, on May 17, he inoculated them again with a more virulent—but still weakened—vaccine. Finally, on May 31, he injected deadly anthrax bacteria into both the vaccinated sheep and 24 sheep who had *not* been vaccinated. Two days later, a large crowd of people—including senators, scientists, and reporters—gathered to witness the dramatic results: All of the vaccinated sheep were alive and well, while those who had not vaccinated were dead or dying.

But perhaps Pasteur's most famous achievement in this area was his creation of a rabies vaccine, his first vaccine for humans. At the time, rabies was a dreaded and almost invariably fatal disease. Typically caught from the bite of an infected dog, treatments of the day ranged from the awful—inserting long heated needles deep into bite wounds—to the horrible—sprinkling gunpowder over the wound and setting it on fire. Although no one knew what caused rabies—the causative virus was too small to see with current microscopes, and it could not be grown in a culture—Pasteur was convinced that the disease was caused by a microbe that attacked the central nervous system. To create his vaccine, Pasteur cultivated the unknown microbe in the brains of rabbits, attenuated it by drying the tissue fragments, and used the fragments in a vaccine.

Although initially reluctant to try the experimental vaccine in humans, on July 6, 1885, Pasteur was compelled to reconsider when nine-year-old Joseph Meister was brought to him with 14 wounds from a rabid dog. Surrendering to the pleas of the boy's mother, Pasteur administered his new vaccine. The lengthy vaccination—13 inoculations over 10 days—was a success, and the boy survived. And despite some public protests that a deadly agent had been inoculated into a human, within 15 months nearly 1,500 others had received the rabies vaccine.

And so, in just eight years, Pasteur not only achieved the first major advance in vaccination since Jenner's time—attenuation—but had also created successful vaccines against chicken cholera, anthrax, and rabies. Yet there was one unexpected twist to his milestone work: It wasn't *all* about reducing the virulence of a virus. As Pasteur himself later realized, most of the viruses in his rabies vaccine were probably not weakened, but *killed*. And therein lay the seeds for the next major milestone.

Milestone #4 A new "killed" vaccine for the birds (not to mention cholera, typhoid, and plague)

By the late 1800s, vaccine development was about to benefit from the birth of a new Golden Age, the discovery of bacteria responsible for numerous diseases, including gonorrhea (1879), pneumonia (1880), typhoid (1880-1884), tuberculosis (1882), and diphtheria (1884). During this time, Theobald Smith, a bacteriologist working at U.S. Department of Agriculture, was assigned to find the microbial culprit responsible for causing hog cholera, a disease that was threatening the livestock industry. While Smith and his supervisor, Daniel Salmon, managed to isolate a causative bacterium, they soon made another far more important discovery: If the microbe was killed by heat and injected into pigeons, the pigeons were then protected against the deadly form of the bacteria. This finding, published in 1886 and soon verified by other researchers, represented a new milestone: Vaccines could be made from *killed*—not merely weakened—cultures of the causative microbe.

The concept of killed vaccines was a major advance in vaccine safety, particularly for those who opposed the idea of vaccines made from live or attenuated microbes. Other scientists soon began trying to make killed vaccines for other diseases, and within just 15 years, the benefactors extended beyond the world of pigeons to the humans affected by three major diseases: cholera, plague, and typhoid.

In the late 1800s, cholera remained a serious problem throughout the world, despite John Snow's milestone work in the late 1840s showing that it was transmitted by contaminated water and Robert Koch's discovery in 1883 that it was caused by a bacterium (*Vibrio cholerae*). While early attempts to create live and attenuated cholera vaccines showed some success, they were abandoned due, in part, to severe reactions. However, in 1896, Wilhelm Kolle achieved a milestone discovery when he developed the first killed vaccine for cholera by exposing cholera bacteria to heat.

Typhoid was another life-threatening disease caused by bacteria (*Salmonella Typhi*) and transmitted by contaminated food or water. While it's still unclear who the first person was to actually inoculate a human with killed typhoid vaccine, in 1896 British bacteriologist Almroth Wright published a paper announcing that a person inoculated with dead salmonella organisms had been successfully protected against the disease. Wright's killed typhoid vaccine was later used with encouraging results in a field trial of 4,000 volunteers from the Indian Army. Tragically, although Wright's vaccine was later used to vaccinate British troops in the Boer War in South Africa, vaccine opponents prevented many others from being vaccinated, going so far as to dump vaccine shipments overboard from transport ships. The result? The British Army suffered more than 58,000 cases of typhoid and 9,000 deaths.

Plague, a disease that killed millions of people in Europe during the Middle Ages, is usually transmitted by bites from fleas carried by rats. The causative bacterium, *Pasteurella pestis* (later renamed *Yersinia pestis*), was discovered 1894. Two years later, Russian scientist Waldemar Haffkine was working in India on a cholera vaccine when the plague broke out in Bombay. Switching his efforts, he soon created a killed vaccine against the plague and, in 1897, tested its safety by inoculating himself. The gamble paid off, and within a few weeks, 8,000 people were vaccinated.

And so by the twentieth century, just a century after Jenner's milestone, the vaccine tally now included two "live" vaccines (smallpox and rabies), three attenuated vaccines (chicken cholera and anthrax), and three killed vaccines (typhoid, cholera, and plague).

Milestone #5 The power of passivity: new vaccines to fight diphtheria and tetanus

In the late 1800s, diphtheria was one of many diseases that took countless human lives, killing as many as 50,000 children a year in Germany alone. Caused by the bacterium *Corynebacterium diptheriae*, diphtheria can cause life-threatening swelling of the airways and damage the heart and nervous system. In 1888, scientists discovered that diphtheria bacteria cause their deadly effects by producing a toxin. Two years later, German physiologist Emil von Behring and Japanese physician Shibasaburo Kitasato made a crucial finding: When animals were infected by diphtheria, they produced a powerful substance that could *neutralize* this toxin—in other words, an *antitoxin*. This finding was followed by another discovery that led to the next milestone in vaccines: If the antitoxin was removed from the animals and injected into *other* animals, it not only protected against diphtheria, but could cure the disease if it was already present.

Just one year later, in December 1891, the first child was inoculated with diphtheria antitoxin and, after further refinements, the vaccine went into commercial production in 1894. Although antitoxin vaccines had their limitations, they would soon be developed against other important diseases, including tetanus.

Antitoxin vaccines were a major advance because they represented a new major concept in vaccines: Active versus passive immunity. Active immunity refers to vaccines that stimulate the body to mount its *own* fight against the microbe, as with the vaccines discussed previously. In contrast, passive immunity involves the transfer of the protective substance from one human or animal to another. Apart from diphtheria and tetanus vaccines, another example of passive immunity is the transfer of antibodies from a mother to her baby during breast feeding. One drawback to passive immunity, however, is that it fades over time, while active immunity is usually permanent.

Von Behring's work in creating a diphtheria vaccine won him the first Nobel Prize in Physiology or Medicine in 1901. But his milestone would soon lead other researchers to solve a greater mystery that had been lurking since Jenner's time: Never mind how they were made—live, attenuated, or killed microbes or antitoxins—exactly *how* did vaccines work?

Milestone #6 An emergence of understanding—and the birth of immunology

Of course, many theories purporting to explain how vaccines might work had been offered over the years. For example, "Exhaustion" theory, held by Pasteur and others, proposed that inoculated microbes consumed "something" in the body until it was depleted and the microbes died off. Another theory, "Noxious Retention," suggested that inoculated microbes produced substances that inhibited their own development. But both theories shared the false view that the body played no active role in vaccines and was merely a passive bystander as the inoculated microbes caused their own demise. Both theories were eventually abandoned in the face of new evidence and new vaccines, and before long the milestone work of two scientists would not only create a new understanding, but a new scientific field and, in 1908, a shared Nobel prize.

A flipped perspective leads to the discovery of the immune system

The roots of Elie Metchnikoff's milestone insight can be traced back to 1883, when the Russian microbiologist performed a landmark experiment in which he observed certain cells have the ability to migrate through the tissues in response to injury or damage. What's more, these cells had the ability to surround, engulf, and digest foreign substances, a process Metchnikoff called phagocytosis (from the Greek *phago*, to devour, and *cytos*, cell). While it initially appeared that cells used phagocytosis as a way to take up nutrients, Metchnikoff suspected they weren't simply out for a Sunday brunch. His hunch was supported by a disagreement he had with Robert Koch, who in 1876 described what he thought were anthrax bacilli invading white blood cells. In his milestone insight, Metchnikoff's flipped this perspective around: The anthrax bacteria were not invading white blood cells; rather, white blood cells were *engulfing and devouring* the bacteria. With this insight, Metchnikoff realized that phagocytosis was a weapon of *defense*, a way to capture and destroy foreign invaders. In short, he had uncovered a cornerstone to the larger mystery of how the body defends itself against disease: the immune system.

By 1887, Metchnikoff had categorized the particle-devouring white blood cells as "macrophages" and, equally important, recognized a key guiding principle by which the immune system operates. In order to function properly, whenever it encounters something in the body, the immune system must "ask" a very basic—but critical—question: Is it "self" or "*non*-self"? If the answer is "non-self"—whether a smallpox virus, anthrax bacterium, or diphtheria toxin—the immune system may begin its attack.

A new theory helps solve the mystery of immunity and how antibodies are made

Like many scientists, Paul Ehrlich's milestone discovery relied in part on new tools that revealed a world previously unseen. For Ehrlich, a German scientist, these tools were dyes—specific chemicals that could be used to stain cells and tissues and thereby reveal new structures and functions. By 1878, when Ehrlich was just 24 years old, they helped him describe several major cells of the immune system, including various types of white blood cells. By 1885, these and other findings led Ehrlich to begin speculating on a new theory of how cells could take in specific nutrients: He proposed that various "side-chains" on the outside of cells—what we now call receptors—could attach to specific substances and bring them inside the cell.

As Ehrlich developed a greater interest in immunology, he began wondering if his receptor theory could explain how diphtheria and tetanus vaccines work. Previously, as we saw, Behring and Kitasato had discovered that when an animal was infected by diphtheria bacteria, it produced an antitoxin and that this antitoxin could be removed and used as a vaccine to protect others against diphtheria. As it turns out, these "antitoxins" were actually antibodies—specific proteins made by cells to target and neutralize the diphtheria toxin. As Ehrlich performed other pioneering work with antibodies, he pondered how his receptor theory might explain how antibodies worked. And he soon arrived at his milestone insight.

While Ehrlich's initial side-chain theory suggested that cells had a large variety of receptors on their outsides, each designed to attach to a specific nutrient, he later expanded this theory and proposed that harmful substances—such as bacteria or viruses—could *mimic* nutrients and also attach to specific receptors. And what happened next, Erhlich proposed, explained how cells produced antibodies against the foreign invader. With the harmful substance attached to its receptor, the cell could then identify key features on the harmful substance and begin producing huge numbers of *new* receptors that were identical to the one that attached to the invader. It was these receptors that then detached from the cell and became antibodies—the highly specific proteins that could seek out, attach to, and inactivate other harmful substances.

Thus, Ehrlich's theory finally explained how specific foreign invaders, once in the body, could be recognized by cells and trigger them to produce specific antibodies that would seek out and attack the invader. The beauty of this theory was that it explained how the body could produce antibodies against specific diseases, whether the antibodies were made as a response to a prior illness, variolation, or vaccination.

Of course, Ehrlich didn't get everything right. For one thing, it turned out that not *all* cells had the ability to attach to foreign invaders and produce antibodies. That critical task was actually accomplished by a specific type of white blood cell—B lymphocytes. What's more, decades of additional research would be required to explain the many complicated roles played by B cells, and many other cells and substances of the immune system.

Nevertheless, today the milestone discoveries of Metchnikoff and Ehrlich are recognized as two complementary cornerstones of modern immunology and of the long-sought explanation of how vaccines work.

Vaccines of the twentieth and twenty-first centuries: the golden age and beyond

By the end of the nineteenth century, vaccines had truly arrived as a major medical breakthrough. Not only had human vaccines been produced for smallpox, rabies, typhoid, cholera, plague, and diphtheria, but most of the fundamental concepts of vaccinology had been introduced. In fact, the subsequent gains made in vaccines throughout the twentieth century can be viewed as refinements to the basic concepts that were known at the end of the nineteenth century.

Nevertheless, vaccines made major advances in the early twentieth century, with the development of vaccines for tuberculosis (1927), yellow fever (1935), pertussis (1926), influenza A (1936), typhus (1938), along with improved vaccines for diphtheria (1923) and tetanus (1927). In addition, in 1931, American pathologist Ernest William Goodpasture introduced a new technique for growing viruses using fertile hen's eggs, resulting in a cheaper and safer way to produce vaccines.

The advances continued after World War II, with the so-called Golden Age of vaccine development. In 1949, John Enders and his associates at Boston Children's Hospital developed a technique for growing viruses in human cells outside a living host; their first efforts not only led to the polio vaccine, but an explosion of vaccine research and advances that continue to this day. Apart from oral and injected polio vaccines, vaccines developed since World War II included those for measles, mumps, rubella, rotavirus, Japanese and tick-borne encephalitis, Lyme disease, hepatitis A and B, meningitis, pneumonia, and influenza, as well as improved vaccines for typhoid, rabies, cholera, typhoid, anthrax, and smallpox.

While the list of recent vaccines is dizzying, a brief look at how they are classified gives a fascinating insight into how vaccines are made today—an astonishing contrast to scratching one's arm with puss from an infected cow.

In the broadest sense, vaccines fall into just two categories: live and inactivated. As we have seen, live, or attenuated, vaccines are made by modifying the disease-producing microbe so it is no longer harmful but still able to stimulate immunity. This category includes both viral and bacterial vaccines, though most live vaccines in the United States today contain attenuated viruses. Today's attenuated viral vaccines include those for measles, mumps, rubella, zoster, rotavirus, and varicella (chickenpox).

Inactivated vaccines include the killed vaccines discussed earlier, as well as several subcategories that truly hint at the complexity and marvel of today's vaccines. The two major types of inactivated vaccines are whole vaccines and fractional vaccines. Whole vaccines are made of either whole or parts of bacteria or viruses and include: 1) viral vaccines against hepatitis A, rabies, and influenza vaccines and 2) bacterial vaccines against pertussis, typhoid, cholera, and plague. Fractional vaccines are where things get interesting. They include three major types: 1) subunit vaccines (made from parts of the disease-causing microbe, such as current vaccines against hepatitis B, influenza, human papillomavirus, and anthrax); 2) toxoid vaccines (modified anti-toxin vaccines, such as improved vaccines against diphtheria and tetanus); and 3) polysaccharide vaccines (made from sugar molecule chains on the surface of certain bacteria, with examples including vaccines against pneumonia and meningitis).

Finally, the new category of recombinant vaccines refers to vaccines made with genetic engineering technology. With genetic engineering, scientists can identify the specific gene in bacteria or viruses that produces a protein that triggers an immune response. The culprit gene is then inserted into yeast cells, which are coaxed to produce large amounts of that protein. The protein is then used to make a vaccine. When the vaccine is administered, it provokes an immune response—that is, causes the body to make antibodies against the protein. Thus, the same antibodies that are produced against the genetically engineered protein will *also* act against a bacteria or virus whose gene originally produced that protein. Genetically engineered vaccines available in the United States include the hepatitis B and human papillomavirus (HPV) vaccines.

The view from today: concerns, transformations, and hope

Today, many health experts consider the discovery of vaccines to be the greatest breakthrough in the history of medicine. They point out, for example, that vaccines have prevented more deaths, permanent disability, and suffering than any other medical discovery or intervention. In fact, some note that with the exception of safe water, no other factor—not even antibiotics—has equaled vaccines in saving human lives.

But apart from the lives saved, vaccines have transformed our lives and view of the world in several profound ways. First, vaccine advances in 1800s contributed significantly to the discovery and acceptance of germ theory—the paradigm-shattering realization that diseases are often caused by invisibly small bacteria and viruses, not evil spirits or religious forces. Second, vaccines opened our eyes to a new world inside our bodies, the immune system, and thus the first true insights into how the body fights disease. Third, vaccines showed us that medicine doesn't always have to involve the blunt force of drugs or surgery. Rather, vaccines *teach* the body to treat itself by inoculating a person with the very disease you're trying to prevent. And finally, vaccines put a new twist on the issue of personal responsibility: with contagious disease, one's decision about whether or not to be vaccinated extends beyond individual health concerns to the health of the entire community.

This last point, whether or not one chooses to be vaccinated, is important and emotionally supercharged by those who resist being "treated" for a disease they don't actually have, fearing that the treatment itself could *cause* the disease. While some concerns about safety are justified, anti-vaccination movements—which have been more or less ongoing since the eighteenth century—can create their own dangers. Arousing fears based on scientifically unfounded claims, such movements often cause people to avoid safe vaccinations and thus increase the risk of epidemic disease.

One recent example is the concern that thimerosal, a mercury-containing preservative used in some vaccines, might cause autism. In 1999, despite a lack of evidence that thimerosal is harmful, the FDA asked pharmaceutical companies to remove the preservative from vaccines. Although many studies have subsequently found no evidence that thimerosal causes neurodevelopmental problems or autism in children, publicity from the ban and the spread of false information by anti-vaccination groups raised sufficient fears among many parents that they stopped vaccinating their children. A 2007 article in the *New England Journal of Medicine* pointed out the dangers of such scenarios with regard to influenza, which each year causes hundreds of thousands of hospitalizations and about 100 deaths of children. Yet, "All of the negative media attention has made many parents reluctant to have their children receive this vaccine..." And thus, the author argues, "By choosing not to vaccinate their children, these parents have elevated a theoretical (and now disproved) risk above the real risk of being hospitalized or killed by influenza."

While the risk of adverse effects is always a legitimate concern, experts note that most well-conducted scientific studies have not supported the idea that vaccines cause rare serious adverse events. In addition, a large body of scientific evidence has discounted purported associations between vaccines and such diseases as multiple sclerosis, measles, mumps, and rubella. As health experts often point out, avoiding vaccination can pose real dangers to the larger population because of the so-called "herd effect," which refers to the fact that the more people that are vaccinated, the better the overall population is protected. Conversely, those who refuse vaccination create gaps in the community defense, offering microbes a free ride to continue their contagious spread.

Apart from safety concerns, real or imagined, vaccines continue to offer an exciting potential for new and better advances in the future. Currently, more than two dozen infections can be prevented by vaccines, and new technologies and strategies, such as those involving the manipulation of genes and proteins, are likely to result in vaccines for many others. Nevertheless, the scientific challenges are many and daunting, as seen with the ongoing quest to find vaccines for malaria and AIDS.

Africa: 16,000 years ago...and today

On October 26, 1977, a hospital cook in Merka, Somalia, became a hero of mixed blessing when he became the last known person on the planet to be infected by smallpox—16,000 years after the virus first made its pathogenic leap in Africa from animal to man. Thus, when the World Health Assembly officially declared the global eradication of smallpox in

1980, it was yet one more remarkable milestone: Smallpox had become the first and only human disease to be eliminated from the planet.

Which makes the announcement that came 30 years later all the more curious. In 2007, the FDA approved a *new* vaccine against... You guessed it, smallpox.

Why a new vaccine against an eradicated disease? The cynical answer is that there will always remain one deadly threat that humanity cannot fully eradicate: itself. Because stockpiled smallpox virus continues to exist for research purposes, new and better vaccines will always be needed to protect us against those who would steal the virus and use it as a weapon against their own species.

And so the battle continues. Viruses will slip into the body on the aerosolized blast of a sneeze and wage their deadly attacks. White blood cells will wage their potent counterattacks with their freshly minted antibodies. And humans will fight each other with whatever sinister weapon they can find. But in the end, at least on one battlefield, vaccines will continue to provide effective ways to help the body mount its valiant—and often victorious—defense.

From Ancient Molds to Modern Miracles: 7

THE DISCOVERY OF ANTIBIOTICS

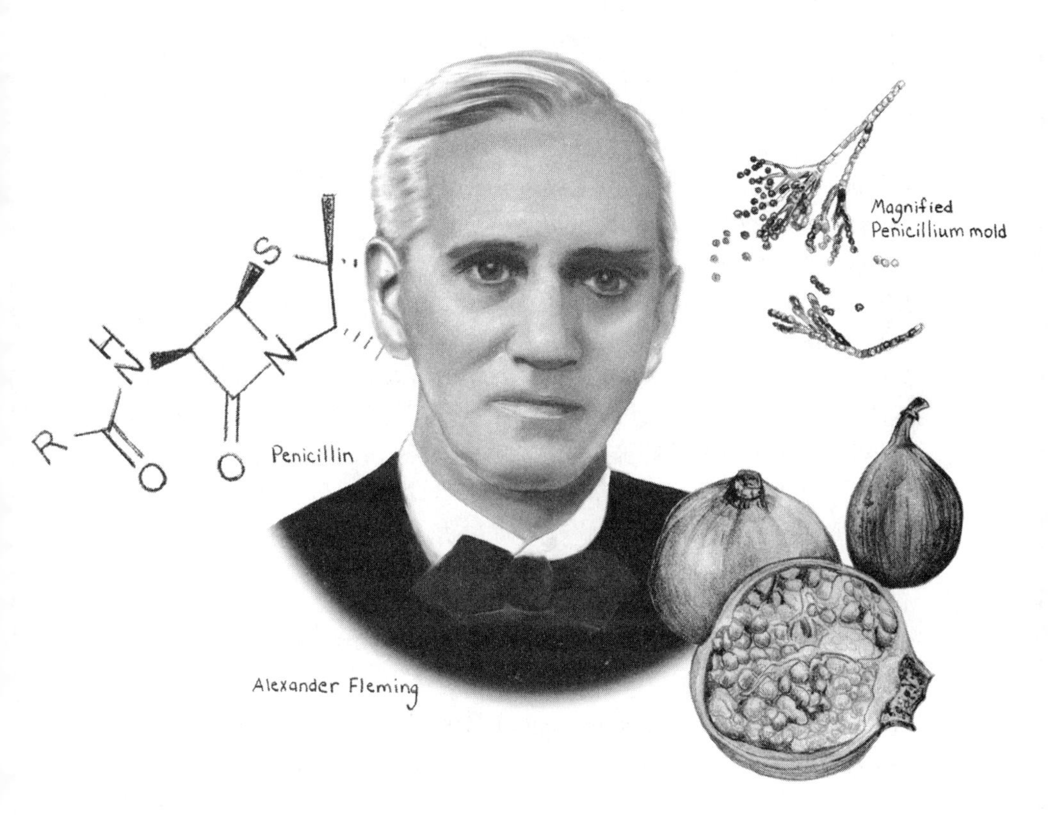

To the villagers who had lived and worked on its slopes for centuries, the 3,000-foot mountain must have seemed idyllic in its beauty and bounty of food that sprang from its rich, fertile soil. Rising from the Bay of Naples along the southwest coast of Italy, its slopes were covered by vineyards, cereal crops, and fruit orchards. Stands of oaks and beech trees rose to its summit and were populated by deer and wild boar. And grazing land was plentiful for the goats that provided milk and cheese. Given that there had been no sign of trouble for more than 1,000 years, it's no wonder that inhabitants of the two towns on its western and southeastern sides—Pompeii and Herculaneum—had no idea that all of this bucolic splendor sat atop a smoldering time bomb, a volcano that would one day erupt with a horrible and fatal fury.

It was on the morning of August 24, AD 79, that Mount Vesuvius, after only a few warning tremors earlier in the month, awoke with sudden violence, sending an enormous "black and terrible cloud" of toxic gas, ash, and cinder ten miles into the sky. During the afternoon, the dark cloud moved southeastward toward Pompeii and began raining down volcanic debris. By the end of the day, the town was covered by a three-foot-deep blanket of ash. While many residents fled in fear, others remained behind, hiding under roofs for shelter. They would meet their fate at around 6 the next morning, when subsequent eruptions overwhelmed the town with burning dust, cinder, and gases, killing as many as 2,000 of its 20,000 residents.

But by that time, the smaller town of Herculaneum—less than 10 miles away on the other side of Vesuvius—had already been devastated. Just hours earlier, at around 1 a.m., an enormous explosion had sent a blast of dense volcanic debris rushing down the western slope at more than 150 miles per hour. Within seconds, Herculaneum was smothered under a 100-foot-deep inferno of ash and cinder. Although most of Herculaneum's 5,000 residents managed to escape in the preceding hours, it was not until 2,000 years later—in 1982—that archeologists excavating a nearby ancient beach discovered the skeletons of 250 individuals who, in their attempt to flee, had not been so lucky. The skeletons were found in various poses on the beach and inside nearby boat sheds and, due to the unusual circumstances of their deaths—instant burial beneath a fine volcanic dust as hot as 1,112 degrees Fahrenheit—were nearly perfectly preserved.

Take two figs and call me in the morning?

Since modern archaeologists began exploring ancient Herculaneum in the 1980s, it's no surprise that they have uncovered a wealth of insights into the lives of these ancient Romans in the days leading up to their deaths. The findings ranged from well-preserved wooden chests and cupboards, to a variety of food remnants, including olive oil, plum jam, dried almonds and walnuts, goat cheese, hardboiled eggs, wine, bread, dried figs, and pomegranates. Nor is it surprising that with modern scientific tools, researchers have been able to learn some remarkable details about the health and diseases suffered by the people whose skeletons were recovered on that ancient beach, including skull lesions caused by scratching lice, rib damage from continuous inhalation of indoor cooking smoke, and foot injuries from Roman shoes and sandals.

What *was* surprising, however, is what the scientists *didn't* find: Evidence of infections.

As described in a 2007 article in the *International Journal of Osteoarchaeology*, in a study of 162 of the 250 Herculaneum skeletons, only *one* showed evidence of general infections. This finding was a "true enigma" because such infections are usually *more* common in ancient populations due to the poor sanitary conditions of the time.

Why were infections so rare among the ancient inhabitants of Herculaneum? A closer look at the villagers' diet uncovered a key clue: Microscopic examination of two particular foods—dried pomegranates and figs—revealed that the fruit was contaminated by *Streptomyces* bacteria. *Streptomyces* are a large and widespread group of generally harmless bacteria that are fascinating for several reasons. For one thing, they are abundant in the soil, where they release a variety of substances that play a critical role in the environment by decomposing plant and animal matter into soil. But equally significant, *Streptomyces* are known today for their ability to produce an astonishing variety of drugs, including up to *two-thirds* of the antibiotics now used in human and veterinary medicine. And one of these antibiotics, tetracycline, is commonly used today to treat a variety of general infections, including pneumonia, acne, urinary tract infections, and the infection that causes stomach ulcers (*Helicobacter pylori*).

Sure enough, when scientists tested the bones of the Herculaneum residents, they found clear evidence that they'd been exposed to the antibiotic tetracycline. Could the villagers have ingested the tetracycline by eating *Streptomyces*-contaminated fruit? In fact, the researchers found that their pomegranates and figs were "invariably contaminated" by the bacteria, probably due to the Roman method of preservation, in which fruit was dried by burying it in beds of straw. That solved one mystery. By eating *Streptomyces*-contaminated pomegranates and figs, the ancient villagers unknowingly dosed themselves with tetracycline antibiotics and thereby protected themselves from general infections. Yet that immediately raises another question: Just how "accidental" was this treatment?

According to historical records, at the same time in history in other parts of the Roman Empire, ancient physicians prescribed a variety of foods to treat infections—including figs and pomegranates. For example, in the first century AD, physician Aulus Cornelius Celsus used pomegranates to treat tonsillitis, mouth ulcers, and other infections, and other Roman physicians used figs to treat pneumonia, gingivitis, tonsillitis, and skin infections. While there's no hard evidence that the physicians of ancient Herculaneum intentionally "prescribed" bacteria-laden fruit to treat infections, it does raise the question: Might these findings cast a new light on exactly who discovered the "first" antibiotic?

* * *

Medical historians need not fear: Neither dried fruits, Ancient Romans, nor *Streptomyces* are likely to usurp the credit traditionally given to three individuals who, 2,000 years later, were awarded the 1945 Nobel Prize in Physiology or Medicine for discovering the first antibiotic: penicillin. The award is well-deserved because when penicillin was first discovered by Alexander Fleming in 1928, and later refined by Howard Florey and Ernst Chain into a more potent version for widespread use, the impact was profound: transforming commonly fatal infections into easily treated conditions and saving millions of lives. Antibiotics, the general term for drugs that inhibit or kill bacteria while leaving normal cells unharmed, became the classic "miracle drug" of the twentieth century and one of the greatest breakthroughs in the history of medicine.

But the story of antibiotics is not without its share of irony and controversy. In the late-nineteenth century, the discovery that bacteria could cause dangerous diseases prompted scientists to search for antibiotics that could fight those diseases. Today—a victim of our own success—the *over-use* of antibiotics is forcing scientists to search again for new antibiotics to treat the *same* diseases.

Setting the stage: from ancient healers to a war among the microbes

To many people, the story of how Alexander Fleming discovered penicillin conjures up a yucky image of mold, the microscopic fungi that make its unwelcome appearance as dark green splotches on a damp shower curtain, old carpet, or loaf of bread. While it's true that many antibiotics—including penicillin—are produced by molds, Fleming did not discover his unique mold in a breadbox or dank bathroom, but on a glass culture plate in his laboratory. Nevertheless, it's fitting that the first known antibiotic was produced by a mold, given that the curative powers of these fuzzy fungi have been recognized by healers and physicians throughout history and in all cultures.

The first written reference to the healing power of mold can be found in the oldest known medical document, a papyrus attributed to the Egyptian healer Imhotep and dating back to around the thirtieth century BC. In that ancient writing, healers are advised that when treating open wounds, they should apply a dressing of fresh meat, honey, oil—and "moldy bread." Later historical accounts report that holy men in Central Asia once applied moldy preparations of "chewed barley and apple" to surface wounds and that in parts of Canada, a spoonful of moldy jam was once commonly ingested to cure respiratory infections. More recently, a physician reported in the 1940s the "well-known fact" that farmers in some parts of Europe once kept loaves of moldy bread handy to treat family members who had been injured by a cut or bruise. The physician wrote that "A thin slice from the outside of the loaf was cut off, mixed into a paste with water, and applied to the wound with a bandage. No infection would then result from such a cut."

But despite such stories, the therapeutic use of mold in folk medicine had no role in the modern discovery of antibiotics. In fact, it was not until the late 1800s that scientists, intrigued by the discovery of bacteria and "germ theory," began to ponder whether it might be possible to cure disease by using one type of microbe to fight another.

One of the earliest reports was by Joseph Lister, the physician who first used antiseptics to prevent surgical infections. In 1871, Lister was experimenting with a species of mold called *Penicillium glaucum*—related to, but not the same as, the mold that led to the discovery of penicillin—when he made an odd discovery: In the presence of this mold, bacteria that normally scurried busily about in every direction across his microscope slide were not only "comparatively languid," but "multitudes were entirely motionless." Lister was so intrigued that he hinted to his brother in a letter that he might investigate whether the mold had a similar effect in people. "Should a suitable case present itself," Lister wrote, "I shall employ *Penicillium glaucum* and observe if the growth of the organisms be inhibited in human tissues." But tantalizingly close as he was to being the first person to discover antibiotics, Lister's investigations did not go far.

A few years later, in 1874, English physician William Roberts made a similar observation in the *Royal Society Note*, pointing out that he was having difficulty growing bacteria in the presence of the same mold. "It seemed," Roberts wrote, "as if this fungus… held in check the growth of bacteria." Two years later, physicist John Tyndall described the antagonistic relationship between *Penicillium* and bacteria in more colorful terms. "The most extraordinary cases of fighting and conquering between bacteria and *Penicillium* have been revealed to me," wrote Tyndall. But Tyndall missed his chance at fame when, instead of investigating whether *Penicillium* was attacking the bacteria through the release of some substance, he mistakenly believed the mold was simply suffocating the bacteria.

Soon, other scientists were making similar observations: To their surprise, the tiny, silent world of microbes was in reality a tumultuous landscape wracked by warfare, not only between molds and bacteria, but between different species of bacteria. In 1889, French scientist Paul Vuillemin was sufficiently impressed by these battles to coin a new term that would foreshadow the coming breakthrough: *antibiosis* ("against life").

Given these early and intriguing clues, why was it not until 1928—another three *decades*—that Fleming finally discovered the first antibiotic? Historians note that several factors may have distracted scientists from pursuing drugs that could fight infections. For one thing, in the late 1800s and early 1900s, physicians were enamored by other recent medical breakthroughs, including antiseptics (chemicals that can kill bacteria on the *surface* of the body but can't be taken internally) and vaccines. What's more, scientists' knowledge of fungi in the nineteenth century was not necessarily to be trusted. In fact, in the early studies of bacteria-fighting fungi, the experimenters might have been referring to any species of *Penicillium* mold—or, for that matter, any green fungus.

And as it turned out, the *Penicillium* mold that led to the discovery of antibiotics was not any old fungus growing on your bathroom wall. It was a specific and rare strain, and the antibiotic substance it produced, penicillin, was fragile, difficult to isolate, and—to be blunt—it was a miracle that Fleming discovered it at all.

Milestone #1 "That's funny": the strange and serendipitous discovery of penicillin

Although most of us prefer to not think about it, just as we are surrounded by countless bacteria, we are similarly exposed to numerous invisible mold spores that waft in daily through our windows and doors, seeking moist surfaces on which to land and germinate. This was pretty much what Alexander Fleming was thinking when, in the summer of 1928, he returned from a long holiday and noticed something growing on a glass culture plate he had left in a corner of his laboratory bench. Fleming was a physician and bacteriologist in the Inoculation Department at St. Mary's Hospital in London, and he had previously inoculated the culture plate with staphylococcus bacteria as part of a research project. Returning from vacation, Fleming had randomly grabbed the glass plate, removed its lid, and was about to casually show it to a colleague, when he peered inside and said, "That's funny…"

Fleming wasn't surprised to see that the surface of the plate was speckled by dozens of colonies of staphylococcus bacteria—that was part of his experiment. Nor was he surprised that one side of the plate was covered by a large splotch of mold. After all, he had been away for two weeks and was planning to discard the culture plate anyway. What caught his eye was something he *didn't* see. Although bacterial colonies covered most of the plate, there was one spot where they came to screeching halt, forming a translucent ring around something they clearly did not like: the giant patch of mold. What's more, the bacteria closest to the mold were clearly disintegrating, as if the mold were releasing something so potent that it was killing them by the millions.

Fortunately, Fleming—who only a few years earlier had discovered lysozyme, a natural bacteria-fighting substance made by a number of tissues in the body—recognized an important discovery when he saw it. As he later wrote, "This was an extraordinary and unexpected appearance, and seemed to demand investigation." Over the next few months, Fleming did exactly that, growing cultures of the mold and studying how the mysterious yellow substance it released affected other types of bacteria. He soon realized that the mold was a specific type of *Penicillium*, and that the substance it released was able to inhibit or kill not just staphylococcus, but many other types of bacteria. A few months later, in 1929, he named the substance "penicillin" and published his first paper about its remarkable properties.

What made penicillin so special? First of all, unlike the lysozyme he had discovered a few years earlier, penicillin stopped or killed many types of bacteria known to cause important human diseases, including staphylococcus, streptococcus, pneumococcus, meningococcus, gonococcus, and diphtheria. What's more, penicillin was incredibly potent. Even in its relatively crude form, it could be diluted to 1 part in *800* before losing its ability to stop staphylococcus bacteria. At the same time, penicillin was remarkably *nontoxic* to the body's cells, including infection-fighting white blood cells.

But apart from penicillin's antibiotic properties, perhaps most amazing was that Fleming discovered it at all. For despite Fleming's own long-held belief, the mold that produced his penicillin did *not* come from a random spore that happened to drift in through an open window in his laboratory and land on his culture plate one summer day. As evidence later showed, the arrival of *that* particular mold spore, the timing of Fleming's vacation, and even local weather patterns all conspired in series of eerie coincidences.

The curious mysteries of a migrant mold

The mystery first came to light when, decades later, a scientist who worked in Fleming's department in the late 1920s recalled that the windows of the laboratory in which Fleming worked were usually *closed*—partly to prevent culture dishes often left on the window sill from falling onto the heads of people walking by on the street below.

If the mold spores didn't drift in from outside, where *did* they come from?

As it turns out, the laboratory in which Fleming worked was located one floor above the laboratory of a scientist named C. J. La Touche. La Touche was a mycologist, a fungi specialist, whose "messy" laboratory just happened to include eight strains of *Penicillium* mold, one of which was later found to be identical to Fleming's mold. But with the windows closed, how did La Touche's mold spores find their way upstairs and into Fleming's culture dish? In another bit of unlikely fortune, Fleming's and La Touche's laboratories were connected by a stairwell whose doorways on both levels were almost always open. Thus, the spores from La Touche's laboratory must have found their way up the open stairwell and onto Fleming's culture plate. Not only that, but the spores had to have appeared at the exact moment when Fleming had removed the lids of his culture plates, either while he was inoculating them with staphylococcus bacteria or perhaps when he was inspecting them under a microscope.

But the serendipity of Fleming's discovery does not end there. Initially, other scientists were unable to repeat Fleming's experiment when their samples of penicillin, strangely, had *no* effect on staphylococcus bacteria. This mystery was later solved when scientists realized that penicillin can only stop bacteria while they are still *growing*. In other words, penicillin has no effect on bacteria once they're fully developed, and the same is true in the body: Whether in the blood or other tissues, penicillin only works against growing bacteria. Which raises another question: How did Fleming's random mold spores manage to germinate and produce penicillin with the exact timing needed to kill the staphylococcus bacteria while they were still growing?

In 1970, Ronald Hare, a professor of bacteriology at the University of London, proposed an extraordinary yet plausible explanation. Researching the weather and temperature conditions at the time of Fleming's vacation, Hare found that Fleming's culture was probably exposed to the mold spores in late July when the temperature was cool enough for the spores to germinate and produce penicillin. Later, weather records showed that the temperature warmed enough for the staphylococcus colonies to grow, just when mold was sufficiently mature to release its penicillin and kill the nearby and still-growing bacteria. If the temperature patterns had been different, the mold might have released its penicillin too late—after the bacteria had stopped growing—and would have been impervious to its antibiotic effects. And Fleming would have seen nothing "funny" about his culture plate when he returned from vacation.

Finally, how likely was it that the spores that happened to land on Fleming's culture were from a penicillin-producing mold and not some other random fungus? While it might seem fairly likely given that the spores came from the nearby laboratory of a fungi specialist, consider this: In the 1940s, scientists undertook an intensive search to find other molds that were as good as Fleming's mold at producing penicillin. Of the approximately *1,000* mold samples they tested, only three—Fleming's and two others—were high-quality producers of penicillin.

* * *

Alexander Fleming's discovery of penicillin in 1928 is considered the launching point for the revolution in antibiotics—but you'd never know it from the amount of attention it received over the following 10 years. Although some scientists read his 1929 paper and were intrigued, and a few physicians tried it in a handful of patients, penicillin was soon all-but-forgotten. As Fleming later explained, he himself was discouraged by several obstacles. First, penicillin was unstable and could lose its potency within days. Second, Fleming didn't have the chemical knowledge to refine it into a more potent form. Finally, Fleming's clinical interest may have been stifled by his physician peers, who were underwhelmed by the idea of treating their patients with a yellow substance made from a moldy broth. And so Fleming soon abandoned penicillin and turned his attention to other work.

Although it would be nearly a decade before penicillin would be "rediscovered," in the meantime two other milestones occurred. One of these was the first authenticated "cure" with penicillin—a milestone achievement by a physician whose name almost no one knows today.

Milestone #2 Like nobody's business: the first successful (but forgotten) cures

Dr. Cecil George Paine was a student at St. Mary's Hospital when he first became intrigued by a lecture Fleming gave and his 1929 paper on penicillin. A few years later, while working as a hospital pathologist, Paine decided to try it out for himself. Around 1931, he wrote to Fleming and asked for cultures of his *Penicillium* mold and, after Fleming complied, soon produced his own samples of crude penicillin. All he needed now were some patients. As Paine later recounted, "I was friendly with one of the eye men, so I asked if he'd like to try out its effects."

The "eye man," Dr. A. B. Nutt, was an assistant ophthalmic surgeon at the Royal Infirmary and apparently the trusting type. He let Paine administer his penicillin treatment in four newborn babies with ophthalmia neonatorum, an eye infection caused by exposure to bacteria in the birth canal. According to the records, one three-week-old boy had a "copious discharge from the eyes" and one six-day-old girl had eyes that were "full of pus." Paine administered his penicillin and recalled later that, "It worked like a charm!" Three of the babies showed significant improvement within two or three days. What's more, Paine later administered penicillin to a mine worker whose lacerated eye had become infected, and "It cleared up the infection like nobody's business."

But despite these historic first cures, Paine abandoned penicillin when he was transferred to another hospital and began pursuing other career interests. He never published his findings and did not receive credit for his work until much later. When once asked where he placed himself in the history of penicillin, Paine replied regretfully, "Nowhere. A poor fool who didn't see the obvious when it was stuck in front of him... It might have come on to the world a little earlier if I'd had any luck."

But if even Paine had published his findings in the early 1930s, was the world even ready for the idea of an "antibiotic" drug? Many historians don't think so because the concept was too novel. After all, how could a drug kill the bacteria that caused an infection, while *not* harming the patient's *own* cells? The medical world was simply not ready, and might not have been, were it not for another key milestone in 1935.

Milestone #3 Prontosil: a forgotten drug inspires a world-changing breakthrough

With penicillin shelved and forgotten by the early 1930s, scientists were investigating a variety of even stranger candidates they hoped could be used to defeat infections. Indeed, some you would sooner expect to find streaming through the iron pipes of a factory than the blood vessels of a human. But, in fact, the idea of using chemicals to treat disease had proven itself in 1910 when Paul Ehrlich—the scientist whose theory of cell receptors in 1885 helped illuminate the immune system and how vaccines work—applied his knowledge of industrial dyes to develop an arsenic-based drug called Salvarsan. Salvarsan was a big hit: The first effective treatment for syphilis, it soon became the most prescribed drug in the *world*.

But after Salvarsan, and up until the early 1930s, scientists had little luck with the use of chemicals to treat infections. One good example of a bad idea was the attempt to use mercurochrome to treat streptococcal infections. Today, the red liquid antiseptic is applied externally to disinfect skin and surface wounds, but in the 1920s some thought infections could be cured with *intravenous injections* of mercurochrome. Fortunately, not everyone bought into this notion: In 1927, a group of researchers argued that any recovery seen in patients injected with mercurochrome was not due to its antibiotic properties, but the patient's subsequent "formidable constitutional disturbances" and "violent purging and rigors."

The push in the 1930s to find *any* antibiotic compound, industrial chemical or otherwise, was understandable. At that time, before the discovery of antibiotics, many infections could quickly turn deadly, including common streptococcal infections, such as strep throat, scarlet fever, tonsillitis, various skin infections, and puerperal (childbirth) fever. The horror of a spreading and unstoppable infection is easily recalled from the story of Mary Wollstonecraft and her agonizing death in 1797 shortly after childbirth (Chapter 3). But though Semmelweis' work in the 1840s eventually helped reduce the incidence of childbirth fever, streptococcal infections remained common and dangerous, particularly if they spread to the blood.

And so it was in this context that, in 1927, German scientist Gerhard Domagk began working in the laboratory of I. G. Farbenindustrie in search of industrial compounds that could fight streptococcal infections. Finally, on December 20, 1932, after testing many chemicals used in the dye industry, Domagk and his associates came upon a chemical from a group of compounds known as sulphonamides. They tested it in the usual way, injecting a group of mice with fatal doses of streptococci bacteria and then, 90 minutes later, giving half of them the new sulphonamide compound. But what they discovered four days later, on December 24, was most *un*usual. While all of the untreated mice were dead from the streptococci bacteria, all of the sulphonamide-treated mice were still alive.

The miracle new drug—later named Prontosil—was soon famous throughout the world. As scientists soon discovered, unlike any previous drug they'd tested, Prontosil could be taken internally not only to treat streptococcal infections, but also gonorrhea, meningitis, and some staphylococcal infections. Soon, other sulphonamide drugs—commonly called "sulfa drugs"—were found, and though none were as effective as Prontosil, in 1939 Domagk was awarded the Nobel Prize in Physiology or Medicine for his work.

Looking back now at the Nobel presentation speech in honor of Domagk's achievement, one can't help noticing an odd disconnect. True, Domagk was properly credited for the "Thousands upon thousands of human lives... being saved each year by Prontosil and its derivatives." But some of the presenter's words seemed to speak to another, even greater, milestone yet to come, particularly when he referred to "a discovery which means nothing less than a revolution in medicine" and "a new era in the treatment of infectious diseases."

But even though Domagk's milestone would soon be overshadowed by penicillin, Prontosil is now widely recognized for opening the medical world to a new way of thinking: It was possible to create drugs that stopped bacterial infections *without* harming the body. And, in fact, Domagk's discovery later helped prod other scientists to take another look at a drug that had been abandoned a decade earlier. As Alexander Fleming himself once noted, "Without Domagk, no sulphonamides; without sulphonamides, no penicillin; without penicillin, no antibiotics."

Milestone #4 From bedpans to industrial vats: the revolution finally arrives

In the late 1930s, two researchers at Oxford University in England began studying the properties of Fleming's antibiotic discovery—not penicillin, but lysozyme, the natural antibiotic Fleming had found in tears and other body fluids several years before the discovery of penicillin. Although the two researchers, German biochemist Ernst Chain and Australian pathologist Howard Florey, were impressed by lysozyme's ability to dissolve bacterial cell walls, by 1939 they had finished their work and were ready to move on. But before writing up their final paper, Chain figured he should take one final look at the literature, and that was when he came across an obscure paper written by Fleming in 1929. Chain was intrigued by what he read about penicillin—not because he was dreaming of antibiotic miracle drugs, but its unique ability to break down bacterial cell walls.

Chain persuaded Florey that they should take a closer look at penicillin, though that was easier said than done: Could they even *find* a sample of the mold nearly a decade after Fleming had abandoned his experiments? But though Fleming's original mold was long gone, Florey and Chain didn't have to look far for its offspring. By coincidence, another staff member at the school had previously obtained a sample from Fleming and had kept it growing ever since. "I was astounded at my luck," Chain later recalled upon learning of the mold. "Here, and in the same building, right under our noses."

Chain began studying the mold, and by early 1940, he applied his background in biochemistry to achieve something Fleming had been unable to do: He produced a small amount of *concentrated* penicillin. Indeed, compared to the "crude" penicillin that Fleming had given up on—which inhibited bacteria at dilutions of 1 part per 800—Chain's concentrated extract was *1,000 times* more potent and able to stop bacteria in dilutions of 1 part per *million*. And yet, amazingly, it was still nontoxic to the body.

Well aware of the recent success of Prontosil and the new hope that drugs could be used to treat infections, Chain and Florey quickly changed their research goals. Penicillin was no longer an abstract curiosity in their study of bacterial cell walls, but a potent *antibiotic*, a therapeutic drug that might be used to cure human diseases. Excitement began to build as Chain and Florey made plans to test their newly potent penicillin in animals. On May 25, 1940, eight mice were given a lethal dose of streptococci, after which four of the mice were given penicillin. The researchers were so excited that they stayed up all night, and by 3:45 a.m., they had their answer: All of the untreated mice were dead, while the mice that received penicillin were still alive.

But once again the researchers were confronted with an obstacle: It had taken Chain considerable time and effort to produce the tiny quantities needed to treat four mice; how could they possibly make enough penicillin for humans? Focusing on the immediate goal of treating just a few people in a clinical trial, research associate Norman Heatley soon found a creative solution. He procured bedpans—hundreds of them—in which the mold could be grown and then used the silk from old parachutes—suspended from a library bookcase—to drain and filter the moldy broth. Chain then chemically extracted penicillin using the methods he'd developed. By early 1941, they had enough penicillin to treat six patients severely sickened by staphylococcus or streptococcus infections. The researchers gave five of the patients intravenous penicillin and one (an infant) oral penicillin. Although one patient eventually died, the other five responded dramatically.

But once again the researchers' excitement was tempered by a seemingly insurmountable challenge: How could they now produce enough penicillin for a larger trial, much less thousands of patients worldwide? By this time, mid-1941, word about the initial trials of penicillin was spreading quickly. It wasn't just that a new antibiotic had been discovered, but that penicillin seemed much more promising than Prontosil and other sulphonamides. As *The Lancet* pointed out in August, 1941, penicillin had a "great advantage" over Prontosil because it not only could fight a greater variety pathogenic bacteria, but it was not affected by pus, blood, or other microbes—*exactly* what you needed in a drug for treating infected wounds.

Yet, given the production limits of bedpans and old parachutes, Florey and Chain still had to figure out how to make large quantities of penicillin. Unfortunately, British pharmaceutical companies were unable to help, as their resources were "stretched to the limit" by Britain's involvement in World War II. And so in June, 1941, Florey and Heatley set out for the United States to seek help from American government and business. Within six months, thanks to good connections and good fortune, Heatley found himself in a laboratory in Peoria, Illinois. Not just any laboratory, but a Department of Agriculture fermentation research laboratory, with the capacity to "brew" an estimated 53,000 gallons of mold filtrate. While hardly enough to treat thousands of patients—100 was closer to the mark—it was far better than the three gallons per hour Heatley had been making back in England.

At this time serendipity appeared twice more in the story of penicillin. First, the Peoria researchers found they could increase penicillin production by about ten times if they augmented the fermentation process with corn steep—a byproduct of cornstarch production that, at the time, could only have been found in the Peoria facility. Then, in another stroke of good fortune, a worker found some mold that, by chance, was growing on a rotting cantaloupe and that produced *six times* more penicillin than Fleming's mold.

With fate and fortune working their magic on both sides of the Atlantic, pharmaceutical companies in America and Britain were soon producing enough penicillin to treat soldiers wounded in World War II—from simple surface wounds to life-threatening amputations. In fact, the increased production was remarkable. In March, 1942, there was barely enough penicillin to treat a single patient; by the end of 1942, 90 patients had been treated; by August, 1943, 500 patients had been treated; and by 1944, thanks to "submerged fermentation" technology developed by Pfizer, there was enough penicillin to treat all of the soldiers wounded in the Invasion of Normandy, as well as a limited number of U.S. civilians.

The discovery of antibiotics—*and* the antibiotic revolution—had finally arrived. But who was the first patient in the United States to actually be saved by penicillin?

Milestone #5

"Black magic": the first patient to be saved by penicillin

In March, 1942, 33-year-old Anne Miller lay dying in Yale-New Haven hospital from a serious streptococcal infection that had spread through her body following a miscarriage. For the past month, doctors had tried and failed to cure her with drugs, surgery, and transfusions. Now, as her condition began to worsen, Miller was slipping in and out of consciousness, and she was not expected to live much longer. That was when her private physician, Dr. John Bumstead, came up with an idea that might just save her life.

Bumstead had been reading about a new drug for treating bacterial infections. He was well aware that only tiny quantities of penicillin had been produced so far, but he also knew one other critical piece of information: Another physician in the hospital, Dr. John Fulton, happened to be an old Oxford friend of one of the few people in the world who might be able to obtain some of the drug: Howard Florey. Oddly enough, Fulton himself lay in a nearby hospital bed, suffering from a severe pulmonary infection. Nevertheless, intent on saving his patient, Bumstead approached the ailing physician and asked if he could somehow persuade

Florey to send some of the rare drug. Despite his feeble condition, Fulton agreed and began working the phone from his hospital bed. Persistence and patience paid off, and on Saturday, March 14, a small package arrived in the mail. Inside was a vial containing a pungent, brown-red powder.

As a group of physicians gathered around the small quantity of penicillin, it wasn't clear exactly *what* they should do with it. After some discussion, they decided to dissolve it in saline and pass it through a filter to sterilize it. Then, bringing it to the bedside of the dying Anne Miller, they gave her an IV injection of 5,000 units. A medical student then gave her subsequent doses every four hours. Prior to that first treatment on Saturday, Miller's temperature was spiking close to 106 degrees Fahrenheit. But with penicillin now coursing through her blood, the effect was immediate and dramatic: Overnight, Miller's temperature dropped sharply. By Monday morning, her fever had plunged to 100 degrees, and she was eating hearty meals. When doctors came to her bed for morning rounds, one senior consultant looked at her temperature chart and was heard to mumble, "Black magic..."

Miller's treatment continued for several months, until her temperature stabilized and she was released. Rescued from her death bed, Miller lived another 57 years, finally passing away in 1999 at the age of 90. While the physicians who treated her in 1942 couldn't know that this remarkable new drug would enable her to live so long, Miller's recovery did have one immediate impact. News of her recovery helped inspire American pharmaceutical companies to dramatically ramp up their production of penicillin, from 400 million units in the first five months of 1943 to 20.5 *billion* units in the next seven months—an increase of 500 times. By 1945, penicillin was being churned out at a rate of 650 billion units a month.

* * *

While the discovery of penicillin, from moldy culture plates in England to giant fermentation vats in Peoria, had its share of serendipity, calculated hard work also played a role in the discovery of antibiotics. In particular, the focused efforts of two highly diligent organisms, bacteria and humans, led medicine into the *next* era—an era where antibiotics seemed to literally—in fact, *did* literally—spring from the ground.

Milestone #6 The battle in the soil: discovery of the second antibiotic (and third and fourth and...)

Dirt: Is there anything simpler, cheaper, or more ubiquitous? We sweep, scrape, and wash it away with a disdain for something whose value is so negligible that it has become the measly standard against which all "dirt cheap" things are compared. Yet to Selman Waksman, dirt held a fascination that dated back to 1915, when he became a research assistant in soil bacteriology at the New Jersey Agricultural Experiment Station. To Waksman, dirt was nothing less than a vast universe populated by a wealth of richly important inhabitants.

Waksman wasn't just interested in the role that microscopic bacteria and fungi play in breaking down plant and animal matter into the organic humus that enables plants to grow. Rather, it was the battle that microbes in the soil continually wage against each *other* and the chemical weaponry they produced to wage those wars. Scientists had known about this microbial warfare for years—it was, you'll recall, why Vuillemin coined the term "antibiosis" in 1889. But what intrigued Waksman was not just that bacteria were continually fighting each other, but that previous findings had shown that *something* in the soil was able to kill one specific bacteria: tubercle bacillus, the bacteria that causes tuberculosis. By 1932, Waksman had shown that whatever that "something" was, it seemed to be released from other bacteria during their ongoing warfare in the soil.

And so in 1939, as other scientists across the Atlantic were taking a second look at a penicillin-inducing mold, Waksman and his associates at Rutgers University in New Jersey began studying dirt and its microbes, hoping that one of them might produce a substance useful in fighting tuberculosis and other human infections. But unlike Fleming's work, serendipity would play no role in Waksman's laboratory. Focusing on a large group of bacteria known as actinomycetes, Waksman's team began a rigorous and systematic investigation in which they looked at more than *10,000* different soil microbes. Their efforts were soon rewarded with the discovery of two substances that had antibiotic properties—actinomycin in 1940 and streptothricin in 1942. While both turned out to be too toxic for use in humans, in September, 1943, Albert Schatz, a PhD student in Waksman's lab, hit "pay dirt" when he found two strains of *Streptomyces* bacteria that produced a substance that could stop other bacteria cold. And not just any bacteria, but tubercle bacillus—the microbe that caused tuberculosis.

The new antibiotic was named streptomycin, and in November, 1943, just weeks after Schatz's discovery, Corwin Hinshaw, a physician at the Mayo Clinic, requested a sample to test in animals. It took five months to get the sample, and it was barely enough to treat four guinea pigs, but it was well worth the wait: The effect of streptomycin on tuberculosis was "marked and striking." Now all Hinshaw needed was a guinea pig of a different sort.

In July, 1943, as she was being admitted to the Mineral Springs Sanitarium in Goodhue County, Minnesota, 19-year-old Patricia Thomas made a confession to her physician: She often spent time with a cousin who had tuberculosis. Her physician was hardly surprised. Thomas herself had now been diagnosed with "far-advanced" tuberculosis, and was deteriorating rapidly. Over the next 15 months, a cavity would form in her right lung, an "ominous" new lesion would appear in her left lung, and she would develop a worsening cough, night sweats, chills, and fever. And so, on November 20, 1944, a year after his success with the rodent, Dr. Hinshaw asked Thomas if she was willing to act as a guinea pig of another sort and become the first human with tuberculosis to be treated with streptomycin. Thomas accepted, and it turned out to be a wise decision. Within six months, she had a rapid—some would say miraculous—recovery. Treatment was stopped in April, 1945, and subsequent X-rays showed significant clearing of disease. Rescued from impending death, Thomas eventually married and gave birth to three children.

Although streptomycin proved to be far from perfect, it was a major milestone in the development of antibiotics. For one thing, like penicillin, streptomycin could fight bacteria in the presence of pus or body fluids. But equally important, streptomycin gave physicians a tool they had never had before: the first effective treatment for tuberculosis. As noted when Waksman received the 1952 Nobel Prize for the discovery, streptomycin had a "sensational" effect on two frequently deadly forms of tuberculosis. In the case of tuberculosis meningitis, an "always fatal" form in which TB bacteria infect the membrane that covers the brain and spinal cord, streptomycin treatment "can be dramatic… Patients that are unconscious and have a high fever may improve rapidly."

Within a few years, streptomycin was known across the world, quickly emerging from a laboratory curiosity to a pharmaceutical best seller being produced at a rate of more than 55,000 pounds a month. Waksman later wrote that the fast rise of streptomycin was partly due to the success of penicillin between 1941 and 1943. Yet streptomycin was a major milestone in its own right, curing thousands of patients with TB who would not have been helped by penicillin. By the end of the 1950s, streptomycin had lowered the death rates of TB in children in some countries by as much as 90%. And streptomycin was only the beginning. Between 1940 and 1952, Waksman and his associates isolated a number of other antibiotics—including actinomycin (1940), clavacin (1942), streptothricin (1942), grisein (1946), neomycin (1948), fradicin (1951), candicidin (1953), and candidin (1953)—though streptomycin and neomycin proved to be the most useful for treating infections.

Waksman also earned one other claim to fame. In the early 1940s, as scientists were publishing more and more papers about "bacteria-fighting" substances, it occurred to Dr. J. E. Flynn, editor of *Biological Abstracts*, that the world needed a new word to call these substances. Flynn asked several researchers for suggestions and considered such terms as "bacteriostatic" and "antibiotin." But Flynn finally settled on a word Waksman had proposed in 1941 or 1942. "Dr. Waksman's reply came back," Flynn later recalled, "and this was the first time I had ever seen the word used in its present sense… as a noun." First used in *Biological Abstracts* in 1943, Waksman's new word—"antibiotic"—is now one of the most recognized terms in medicine throughout the world.

Antibiotics today: new confidence, new drugs, new concerns

Since their discovery in the 1940s, antibiotics have transformed the world in many ways—good, bad, predictable, and unpredictable. Today, it's difficult to imagine the fear patients must have felt prior to the 1940s, when even minor injuries and common diseases could erupt into fast-spreading deadly infections. With antibiotics, physicians were suddenly empowered with the most satisfying tools imaginable—pills, ointments, and injections that could dramatically save lives.

Yet some also claim that antibiotics have a brought out a darker side of human nature, the mere knowledge of their availability creating a sense of false confidence and a willingness to indulge in risky behaviors.

For example, antibiotics may have helped create a society more focused on the convenience of treatment, than the hard work of prevention. Equally alarming, some claim antibiotics have contributed to an increase in immoral behaviors, as seen in the epidemics of sexually transmitted diseases. Finally, even though antibiotics have saved millions of lives, it's important to remember that they are not available to everyone and do not always work: Each year, as many as 14 million people around the world *still* die from infections.

* * *

Although many antibiotics have been discovered since the 1940s—90 antibiotics were marketed between 1982 and 2002 alone—it's helpful to remember that all antibiotics share one common principle: the ability to stop the infecting microbe without harming the patient's own cells. They achieve this by capitalizing on a vulnerability found in the microbe but *not* in human cells. Based on this principle, antibiotics generally fall into one of four categories:

1. **Folate antagonists**—Prontosil and other sulfa drugs in this category prevent bacteria from synthesizing folic acid, a substance they need to grow and replicate.
2. **Inhibitors of cell wall synthesis**—Penicillin and related drugs in this group prevent bacteria from making their cell walls.
3. **Protein synthesis inhibitors**—Streptomycin, neomycin, tetracycline, and many others in this category target ribosomes (tiny structures inside bacteria that make proteins).
4. **Quinolone antibiotics**—Often used to treat urinary tract infections, these agents inhibit an enzyme bacteria needed to replicate their DNA.

How do physicians choose an antibiotic among the many available drugs? One key factor is whether the infecting bacteria have been identified and are known to be sensitive to a specific antibiotic. Other factors include the location of the infection, patient health (a healthy immune system is needed to help defeat the infection), side effects, and cost. But perhaps the first and most important question is whether an antibiotic should be used at all. Even back in 1946, Fleming cautioned that penicillin has no effect on "cancer, rheumatoid arthritis... multiple sclerosis, Parkinson's disease... psoriasis, and almost all the viral diseases, such as small pox, measles, influenza, and the common cold." If some of these seem laughably obvious, consider that Fleming added, "These are merely some of the diseases of many suffers who, in the past two years, have as a result of press reports written to me for relief."

Unfortunately, improper use still hovers like a dark cloud over one of the ten greatest discoveries in medicine. The issue centers around the emergence of resistance—the ability of bacteria to adapt, survive, and thrive despite treatment with an antibiotic—which can occur when antibiotics are improperly used. Bacteria are notoriously clever at developing resistance, for example by acquiring genetic mutations that protect them against the drug or by producing enzymes that inactivate the drug. As bacteria pass on such traits to new generations, they can evolve into "superbugs" that are resistant to multiple antibiotics and that transform once-treatable diseases into potentially deadly infections. While natural processes play a role in resistance, it's now clear that careless and inappropriate use by humans is a major factor.

Mis-use and neglect: an age-old problem causes a brand-new danger

The warning signs were evident as early as the 1940s, when the presenter of the Nobel Prize to Selman Waksman cautioned that one complication already seen in the treatment of tuberculosis was "the development of strains of bacteria that become more and more resistant to streptomycin…" Other warnings about resistance appeared in the 1950s and early 1960s, when Japanese physicians reported an epidemic of bacterial dysentery that had become resistant to streptomycin, tetracycline, and chloramphenicol. And in 1968, doctors reported the first outbreak of bacterial infections that were resistant to methicillin and other penicillin-type drugs. Since then, those resistant bacteria—called methicillin-resistant *Staphylococcus aureus*, or MRSA—have become a worldwide concern.

S. aureus is a common bacteria found on the skin and normally quite harmless, even when it enters the skin through a cut or sore and causes a local infection such as a pimple or boil. But in people with weakened immune systems, such infections can turn deadly if they spread to the heart, blood, or bone, and particularly if antibiotics lose their effectiveness against them. Unfortunately, this is exactly what began to happen in the 1970s when MRSA began appearing in hospitals and killing as many as 20 to 25% of the people it infected. Worse, in the past decade MRSA has begun to venture outside hospitals, with so called "community-acquired" MRSA outbreaks now occurring in prisons, nursing homes, and school sports teams. A recent article in the *New England Journal of Medicine* found that emerging strains of MRSA are now showing resistance to vancomycin, another important antibiotic. The authors wrote that the problem stems not only from improper use of antibiotics, but is "exacerbated by a dry pipeline" for new antibiotics. They concluded that "a concerted effort on the part of academic researchers and their institutions, industry, and government is crucial if humans are to maintain the upper hand in this battle against bacteria—a fight with global consequences."

Describing the development of new antibiotics as a "dry pipeline" may seem odd given how many have been produced since the 1940s, but as it turns out, most commonly used antibiotics today were discovered in the 1950s and 1960s. Since then, pharmaceutical companies have mostly tweaked them to create new chemical variations. But as one author pointed out in a recent issue of *Biochemical Pharmacology*, "It is still critically important to find new antibiotic classes [given] the increasing incidence of resistant pathogens. If we do not invest heavily in discovering and developing new antibiotic classes, we might well end up in a situation akin to the pre-antibiotic era..."

While some have hoped that biotechnology would lead to revolutionary new antibiotics, so far such technologies have produced limited advances at best. For this reason, other researchers suggest that we indeed may need to return to the "pre-antibiotic" era by taking a harder look at the natural world, the microbes that have been making antibiotics for far longer—a half-billion years or so—than humans.

Overcoming resistance: a way out through the past?

Given that two-thirds of our current antibiotics *already* come from *Streptomyces* bacteria, some might wonder if it really makes sense to continue to investigate "natural resources" for new antibiotics. But in fact, we have barely seen the tip of the iceberg.

How big an iceberg? In a 2001 issue of the *Archives of Microbiology*, researchers made an eye-opening claim: They found that *Streptomyces* bacteria, which include 500 or more separate species, may be capable of producing as many as *294,300* different antibiotics. If you're wondering how a group of one-celled organisms could be so productive, consider the genetic engines packed into these tiny one-celled creatures. In 2002, other researchers announced in *Nature* that they had decoded the entire genetic sequence of a representative species of *Streptomyces*, uncovering an estimated 7,825 genes. This was the largest number of genes found in a bacterium, and nothing to sneer at given that it's about *one-third* the number found in humans. With that kind of genetic abundance, perhaps it's not surprising that these microbial super-specialists, rather than putting their genes to work making multicellular arms, legs, and cerebrums, are capable of producing so many different antibiotics.

* * *

In the early 1980s, anthropologists uncovered the skeletons of an ancient group of people who died more than 1,000 years ago and whose remains were remarkably well-preserved. Conducting fluorescent studies, the scientists found evidence of the antibiotic tetracycline in their bones and postulated that it may have been produced by Streptomycetes *bacteria present in the food that the people ate at the time. The researchers also speculated that the tetracycline in their food might account for the "extremely low rates of infectious diseases" found in these people.*

No—we're not talking about the villagers of Herculaneum in 79 AD, but a group of Sudanese Nubians who lived on the west bank of the Nile River a few hundred years later, in 350 AD. And the source of their dietary tetracycline was not dried pomegranates or figs, but wheat, barley, and millet grains they had stored in mud bins. The scientists speculated that the mud storage bins provided the ideal environment needed for Streptomycetes—which comprise up to 70% of the bacteria in the desert soils of Sudanese Nubia—to flourish. It was unclear whether the tetracycline found in the ancient Nubians was produced by the same species of Streptomyces *that produced the tetracycline found in the people of ancient Herculaneum.*

But that's exactly the point. In an age of emerging resistance and potentially deadly infections, could this remarkable genus of bacteria—provider of antibiotics to ancient peoples, source of nearly a dozen antibiotics discovered in the 1940s and 1950s, producer of two-thirds of the current antibiotics in use today, and yet still barely tapped for their potential of nearly 300,000 antibiotics—could this remarkable genus of bacteria be trying to tell us something?

Breaking God's Code: THE DISCOVERY OF HEREDITY, GENETICS, AND DNA 8

One fine day at the dawn of civilization, on the beautiful Greek island of Kos in the crystal-clear waters of the Aegean Sea, a young noblewoman slipped quietly through the back entrance of a stone and marble healing temple known as the Asklepieion and approached the world's first and most famous physician. In desperate need of advice, she presented Hippocrates with an awkward problem. The woman had recently given birth to a baby boy, and though the infant was plump and healthy, Hippocrates needed only glance from fair-skinned mother to her swaddling infant to make his diagnosis: The baby's dark skin suggested that she had a passionate tryst in the recent past with an African trader. And if news of the infidelity got out, it would spread like wildfire, enraging her husband and inflaming scandalous gossip across the island.

But Hippocrates—as knowledgeable in the science of heredity and genetics as anyone could be in fifth century BC—quickly proposed another explanation. True, some physical traits could be inherited from the father, but that did not take into account the concept of "maternal impression." According to that view, babies could also acquire traits based on what their mothers looked at during pregnancy. And thus, Hippocrates assured her, the baby must have acquired its negroid features during pregnancy, when the woman had gazed a bit too long at a portrait of an Ethiopian that happened to be hanging from her bedroom wall.

From guessing games to a genetic revolution

Since the earliest days of civilization until well after the Industrial Revolution, people from every walk of life struggled valiantly, if not foolishly, to decipher the secrets of heredity. Even today, we *still* marvel at the mystery of how traits are passed from generation to generation. Who has not looked at a child or sibling and tried to puzzle out what trait came from whom—the crooked smile, skin color, intelligence or its lack, the perfectionist or lazy nature? Who has not wondered why a child inherited this from the mother, that from the father, or how brothers or sisters can be *so* different?

And those are just the obvious questions. What about traits that disappear for a generation and reappear in a grandchild? Can parents pass traits that they "acquire" during their lives—a skill, knowledge, even an injury—to their children? What role does environment play? Why are some families haunted for generations by disease, while others are graced with robust health and breathtaking longevity? And perhaps most troubling: What inherited "time bombs" will influence how and when we die?

Up until the twentieth century, all of these mysteries could be summed up with two simple questions: Is heredity controlled by any rules? And how does it even *happen*?

Yet amazingly, despite no understanding of how or why traits are passed from generation to generation, humans have long *manipulated* the mystery. For thousands of years, in deserts, plains, forests, and valleys, early cultures cross-mated plant with plant and animal with animal to create more desirable traits, if not entirely new organisms. Rice, corn, sheep, cows, and horses grew bigger, stronger, hardier, tastier, healthier, friendlier, and more productive. A female horse and a male donkey produced a mule, which was both stronger than its mother *and* smarter than its father. With no understanding of *how* it worked, humans played with heredity and invented agriculture—an abundant and reliable source of food that led to the rise of civilization and transformation of the human race from a handful of nomads into a population of billions.

Only in the past 150 years—really just the past 60—have we begun to figure it out. Not all of it, but enough to decipher the basic laws, to pull apart and poke at the actual "stuff" of heredity and apply the new knowledge in ways that are now on the brink of revolutionizing virtually every branch of medicine. Yet perhaps more than any other breakthrough, it has been a 150-year explosion in slow motion because the discovery of heredity—how DNA, genes, and chromosomes enable traits to be passed from generation to generation—is in many ways still a work in progress.

Even after 1865, when the first milestone experiment revealed that heredity *does* operate by a set of rules, many more milestones were needed—from the discovery of genes and chromosomes in the early 1900s to the discovery of the structure of DNA in the 1950s, when scientists finally began to figure out how it all actually *worked*. It took one and a half centuries to not only understand how traits are passed from parent to child, but how a tiny egg cell with no traits can grow into a 100-trillion-cell adult with many traits.

And we are still only at the beginning. Although the discovery of genetics and DNA was a remarkable breakthrough, it also flung open a Pandora's box of possibilities that boggle the mind—from identifying the genetic causes of diseases, to gene therapies that cure disease, to "personalized" medicine in which treatment is customized based on a person's unique genetic profile. Not to mention the many related revolutions, including the use of DNA to solve crimes, reveal human ancestries, or perhaps one day endow children with the talents of *our* choice.

***

Long after the time of Hippocrates, physicians remained intrigued by the idea of maternal impression, as seen in three case reports from the nineteenth and early twentieth centuries:

- *A woman who was six months pregnant became deeply frightened when, seeing a house burning in the distance, she feared it might be her own. It wasn't, but for the rest of her pregnancy, the terrifying image of flames was "constantly before her eyes." When she gave birth to a girl a few months later, the infant had a red mark on her forehead whose shape bore an uncanny resemblance to a flame.*
- *A pregnant woman became so distressed after seeing a child with an unrepaired cleft lip that she became convinced her own baby would be born with the same trait. Sure enough, eight months later her baby* was *born with a cleft lip. But that's only half the story: A few months later, after news of the incident had spread and several pregnant women came to see the infant,* three more *babies were born with cleft lips.*
- *Another woman was six months pregnant when a neighborhood girl came to stay in her house because her mother had fallen ill. The woman often did housework with the girl, and thus frequently noticed the middle finger on the girl's left hand, which was partly missing due to a laundry accident. The woman later gave birth to a boy who was normal in all ways—except for a missing middle finger on his left hand.*

Shattering the myths of heredity: the curious lack of headless babies

Given how far science has come in the past 150 years, one can understand what our ancestors were up against in their attempts to explain how we inherit traits. For example, Hippocrates believed that during conception, a man and woman contributed "tiny particles" from every part of their bodies, and that the fusion of this material enabled parents to pass traits to their children. But Hippocrates' theory—later called

pangenesis—was soon rejected by the Greek philosophe
part because it didn't explain how traits could *skip* a ge
course, Aristotle had his own peculiar ideas, believing t
received physical traits from the mother's menstrual blood ar
from the father's sperm.

Lacking microscopes or other tools of science, it's no surprise that heredity remained a mystery for more than two millennia. Well into the nineteenth century, most people believed, like Hippocrates, in the "doctrine of maternal impression," the idea that the traits of an unborn child could be influenced by what a woman looked at during pregnancy, particularly shocking or frightening scenes. Hundreds of cases were reported in medical journals and books claiming that pregnant women who had been emotionally distressed by something they had witnessed—often a mutilation or disfigurement—later gave birth to a baby with a similar deformity. But doubts about maternal impression were already arising in the early 1800s. "If shocking sights could produce such effects," asked Scottish medical writer William Buchan in 1809, "how many headless babies would have been born in France during Robespierre's reign of terror?"

Still, many strange myths persisted up until the mid-1800s. For example, it was commonly rumored that men who had lost their limbs to cannon-fire later fathered babies without arms or legs. Another common misconception was that "acquired traits"—skills or knowledge that a person learned during his or her lifetime—could be passed on to a child. One author reported in the late 1830s that a Frenchman who had learned to speak English in a remarkably short time must have inherited his talent from an English-speaking grandmother he had never met.

As for *which* traits come from which parent, one nineteenth century writer confidently explained that a child gets its "locomotive organs" from the father and its "internal or vital organs" from the mother.

This widely accepted view, it should be added, was based on the appearance of the mule.

The first stirrings of insight: microscopes help locate the stage

And so, as late as the mid-1800s, even as scientific advances were laying the groundwork for a revolution in many areas of medicine, heredity continued to be viewed as a fickle force of nature, with little agreement among scientists on where it occurred and certainly no understanding of how it happened.

The earliest stirrings of insight began to arise in the early 1800s, thanks in part to improvements to the microscope. Although it had been 200 years since Dutch lens grinders Hans and Zacharias Janssen made the first crude microscope, by the early 1800s technical improvements were finally enabling scientists to get a better look at the scene of contention: the cell. One major clue came in 1831 when Scottish scientist Robert Brown discovered that many cells contained a tiny, dark central structure, which he called the nucleus. While the central role the cell nucleus played in heredity would not be known for decades, at least Brown had found the stage.

Ten years later, British physician Martin Barry helped set that stage when he realized that fertilization occurred when the male's sperm cell *enters* the female's egg cell. That may sound painfully obvious today, but only a few decades earlier another common myth was that every unfertilized egg contained a tiny "preformed" person, and that the job of the sperm was to poke it to life. What's more, up until the mid-1800s, most people didn't realize that conception involved only *one* sperm and *one* egg. And without knowing that simple equation—1 egg + 1 sperm = 1 baby—it would be impossible to take even the first baby steps toward a true understanding of heredity.

Finally, in 1854, a man came along who was not only aware of that equation, but willing to bet a decade of his life on it to solve a mystery. And though his work may *sound* idyllic—working in the pastoral comfort of a backyard garden—in reality his experiments must have been numbingly tedious. Doing something no one had ever done, or perhaps dared think of doing, he grew tens of thousands of pea plants and painstakingly documented the traits of their little pea plant offspring for generation after generation. As he later wrote with some pride, "It requires indeed some courage to undertake a labor of such far reaching extent."

But by the time Gregor Mendel finished his work in 1865, he had answered a question humanity had been asking for thousands of years: Heredity is not random or fickle, but does *indeed* have rules. And the other fringe benefit—apart from a pantry perpetually stocked with fresh peas? He had founded the science of genetics.

Milestone #1 From peas to principles: Gregor Mendel discovers the rules of heredity

Born in 1822 to peasant farmers in a Moravian village (then a part of Austria), Gregor Mendel was either the most unlikely priest in the history of religion or the most unlikely researcher in the history of science—or perhaps both. There's no question of his intelligence: An outstanding student as a youth, one of Mendel's teachers recommended that he attend an Augustinian Monastery in the nearby city of Brünn, a once-common way for the poor to enter a life of study. Yet by the time Mendel was ordained as a priest in 1847 at the age of 26, he seemed poorly suited for religion *or* academics. According to a report sent to the Bishop of Brünn, at the bedside of the ill and suffering, Mendel "was overcome with a paralyzing shyness, and he himself then became dangerously ill."

Mendel fared no better a few years later when, having tried substitute teaching in local schools, he failed his examination for a teacher's license. To remedy this unfortunate outcome, he was sent to the University of Vienna for four years, studied a broad variety of subjects, and in 1856 took the exam a second time.

Which he promptly failed again.

And so, with dubious credentials as a priest and scholar, Mendel returned to his quiet life at the monastery, perhaps resolved to spend the rest of his life as a humble monk and part-time teacher. Or perhaps not. Because while Mendel's training failed to help him pass his teaching exams, his education—which had included fruit growing, plant anatomy and physiology, and experimental methods—seemed calculated for something of greater interest to him. As we now know, as early as 1854, two years *before* failing his second teacher's exam, Mendel was already conducting experiments in the abbey garden, growing different varieties of pea plants, analyzing their traits, and planning for a far grander experiment that he would undertake in just two years.

Eureka: 20,000 traits add up to a simple ratio and two major laws

What was on Mendel's mind when he began his famous pea plant experiment in 1856? For one thing, the idea did not strike him out of the blue.

As in other areas of the world, cross-mating different breeds of plants and animals had long been of interest to farmers in the Moravian region as they tried to improve ornamental flowers, fruit trees, and the wool of their sheep. While Mendel's experiments may have been partly motivated by a desire to help local growers, he was also clearly intrigued by larger questions about heredity. Yet if he tried to share his ideas with anyone, he must have left them scratching their heads because at the time, scientists didn't think traits were something you could study at *all*. According to the views of development at the time, the traits of plants and animals blended from generation to generation as a *continuous* process; they were not something you could separate and study individually. Thus, the very design of Mendel's experiment—comparing the traits of pea plants across multiple generations—was a queer notion, something no one had thought to do before. And—not coincidentally—a brilliant leap of insight.

But all Mendel was doing was asking the same question countless others had asked before him: Why did some traits—whether Grandpa's shiny bald head or auntie's lovely singing voice—disappear in one generation and reappear in another? Why do some traits come and go randomly, while others, as Mendel put it, reappear with "striking regularity"? To study this, Mendel needed an organism that had two key features: physical traits that could be easily seen and counted; and a fast reproductive cycle so new generations could be produced fairly quickly. As luck would have it, he found it in his own backyard: *Pisum sativum*, the common pea plant. As Mendel began cultivating the plants in the abbey garden in 1856, he focused on seven traits: flower color (purple or white), flower location (on the stem or tip), seed color (yellow or green), seed shape (round or wrinkled), pod color (green or yellow), pod shape (inflated or wrinkled), and stem height (tall or dwarf).

For the next eight years, Mendel grew thousands of plants, categorizing and counting their traits across countless generations of offspring. It was an enormous effort—in the final year alone, he grew as many as *2,500* second-generation plants and, overall, he documented more than *20,000* traits. Although he didn't finish his analysis until 1864, intriguing evidence began to emerge from almost the very start.

To appreciate Mendel's discovery, consider one of his simplest questions: Why was it that when you crossed a purple-flowered pea plant with a white-flowered plant, *all* of their offspring had purple flowers; and yet, when *those* purple-flowered offspring were crossed with each other, most of the offspring in the next generation had purple flowers, while a *few* had white flowers? In other words, where in that first generation of all purple-flowered plants were the "instructions" to make white flowers hiding? The same thing occurred with the other traits: If you crossed a plant with yellow peas with a plant with green peas, *all* of the first-generation plants had yellow peas; yet when those pea plants were crossed with each other, most second-generation offspring had yellow peas, while *some* had green peas. Where, in the first generation, were the instructions to make green peas hiding?

It was not until Mendel had painstakingly documented and categorized thousands of traits over many generations that the astonishing answer began to emerge. Among those second-generation offspring, the same curious ratio kept appearing over and over: *3 to 1*. For every 3 purple-flowered plants, there was 1 white-flowered plant; for every 3 plants with yellow peas, 1 plant with green peas. For every 3 tall plants, 1 dwarf plant, and so on.

To Mendel, this was no statistical fluke, but suggested a meaningful principle, an underlying *Law*. Delving into how such patterns of inheritance could occur, Mendel began envisioning a mathematical *and* physical explanation for how hereditary traits could be passed from parent to offspring in this way. In a remarkable milestone insight, he deduced that heredity must involve the transference of some kind of "element" or factor from each parent to the child—what we now know to be *genes*.

And that was just the beginning. Based on nothing more than his analysis of pea plant traits, Mendel intuited some of the most important and fundamental laws of inheritance. For example, he correctly realized that for any given trait, the offspring must inherit two "elements" (genes)—one from each parent—and that these elements could be *dominant* or *recessive*. Thus for any given trait, if an offspring inherited a dominant "element" from one parent and a recessive "element" from the other, the offspring would show the dominant trait, but also *carry* the *hidden* recessive trait and thus potentially pass it on to the next generation. In the case of flower color, if an offspring had inherited a dominant purple gene from one parent and a recessive white gene from the other, it would have purple-flowers but *carry* the recessive white-flower gene, which it could pass on to its offspring. This, finally, explained how traits could "skip" a generation.

Based on these and other findings, Mendel developed his two famous laws of how the "elements" of heredity are transferred from parent to offspring:

1. The *Law of Segregation:* Although everyone has two copies of every gene (one from each parent), during the production of sex cells (sperm and eggs) the genes separate (segregate) so that each egg and sperm only has one copy. That way, when the egg and sperm join in conception, the fertilized egg again has two copies.
2. The *Law of Independent Assortment:* All of the various traits (genes) a person inherits are transferred *independently* of each other; this is why, if you're a pea plant, just because you inherit a gene for white flowers doesn't mean you will also *always* have wrinkled seeds. It is this "independence" of gene "sorting" that we can thank for the huge variety of traits seen in the human race.

To appreciate the genius of Mendel's accomplishment, it's important to remember that at the time of his work, no one had yet observed *any* physical basis for heredity. There was no conception of DNA, genes, *or* chromosomes. With no knowledge of what the "elements" of heredity might be, Mendel discovered a new branch of science whose defining terms—gene and genetics—would not be coined until decades later.

A familiar theme: confident and unappreciated to his death

In 1865, after eight years of growing thousands of pea plants and analyzing their traits, Gregor Mendel presented his findings to the Brünn Natural Historical Society, and the following year, his classic paper, "Experiments in Plant Hybridization," was published. It was one of the greatest milestones in the history of science *or* medicine and answered a question mankind had been asking for millennia.

And the reaction? A rousing yawn.

Indeed, for the next 34 years Mendel's work was ignored, forgotten, or misunderstood. It wasn't that he didn't try: At one point, he sent his paper to Carl Nägeli, an influential botanist in Munich. However, Nägeli not only failed to appreciate Mendel's work, but wrote back with perhaps one of the most galling criticisms in the annals of science. Having perused a paper based on nearly *ten years* of work and growing more than *10,000* plants, Nägeli wrote, "It appears to me that the experiments are not completed, but that they should really just begin…"

The problem, historians now believe, was that Mendel's peers simply could not grasp the significance of his discovery. With their fixed views of development and the belief that hereditary traits could not be separated and analyzed, Mendel's experiment fell on deaf ears. Although Mendel continued his scientific work for several years, he finally stopped around 1871, shortly after being named abbot of the Brünn monastery. When he died in 1884, he had no clue that he would one day be known as the founder of genetics.

Nevertheless, Mendel was convinced of the importance of his discovery. According to one abbot, months before his death Mendel had confidently said, "The time will come when the validity of the laws discovered by me will be recognized." Mendel also reportedly told some monastery novices shortly before he died, "I am convinced that the entire world will recognize the results of these studies."

Three decades later, when the world finally *did* recognize his work, scientists discovered something else Mendel didn't know, but that put his work into final, satisfying, perspective. His laws of inheritance applied not only to plants, but to animals *and* people.

But if the science of genetics had arrived, the question now was, *where*?

Milestone #2

Setting the stage: a deep dive into the secrets of the cell

Although the next major milestone began to emerge in the 1870s, about the same time Mendel was giving up on his experiments, scientists had been laying the foundation for centuries before then. In the 1660s, English physicist Robert Hooke became the first person to peer through a crude microscope at a slice of cork bark and discover what he referred to as tiny "boxes." But it was not until the 1800s that a series of German scientists were able to look more closely at those boxes and finally discover where heredity is played out: the cell and its nucleus.

The first key advance occurred in 1838 and 1839 when improvements to the microscope enabled German scientists Matthias Schleiden and Theodor Schwann to identify cells as *the* structural and functional units of all living things. Then in 1855, dismissing the myth that cells appeared out of nowhere through spontaneous generation, German physician Rudolf Virchow announced his famous maxim, *Omnis cellula e cellula*—"Every cell comes from a pre-existing cell." With that assertion, Virchow provided the next key clue to *where* heredity must happen: If every cell came from another cell, then the information needed to make each new cell—its *hereditary* information—must reside somewhere *in* the cell. Finally, in 1866, German biologist Ernst Haeckel came right out and said it: The transmission of hereditary traits involved something in the cell *nucleus*, the tiny structure whose importance was recognized in 1831 by Robert Brown.

By the 1870s, scientists were diving deeper and deeper into the cell nucleus and discovering the mysterious activities that occurred whenever a cell was about to divide. In particular, between 1874 and 1891, German anatomist Walther Flemming provided detailed descriptions of these activities, which he called mitosis. Then in 1882, Flemming gave the first accurate descriptions of something peculiar that happened just before a cell was about to divide: long threadlike structures became visible in the nucleus and split into two copies. In 1888, as scientists speculated on the role those threads might play in heredity, German anatomist Wilhelm Waldeyer—one of the great "namers" in biology—came up with a term for them, and it stuck: *chromosomes*.

Milestone #3

The discovery—and dismissal—of DNA

As the nineteenth century drew to a close, the world, already busy ignoring the first great milestone in genetics, went on to dismiss the *second* great milestone: the discovery of DNA. That's right, *the* DNA, the substance from which genes, chromosomes, hereditary traits and, for that matter, the twenty-first century revolution in genetics are built. And, like the snubbing of Mendel and his laws of heredity, it was no short-term oversight: Not long after its discovery in 1869, DNA was essentially put aside for the next half century.

It all began when Swiss physician Friedrich Miescher, barely out of medical school, made a key career decision. Because his poor hearing, damaged by a childhood infection, made it difficult to understand his patients, he decided to abandon a career in clinical medicine. Joining a laboratory at the University of Tübingen, Germany, Miescher decided instead to look into Ernst Haeckel's recent proposal that the secrets of heredity might be found in the cell nucleus. That might sound like a glamorous endeavor, but Miescher's approach was decidedly not. Having identified the best type of cell for studying the nucleus, he began collecting dead white blood cells—otherwise known as pus—from surgical bandages freshly discarded by the nearby university hospital. Wisely, he rejected any that were so decomposed that they, well, stunk.

Working with the least-offensive samples he could find, Miescher subjected the white blood cells to a variety of chemicals and techniques until he succeeded in separating the tiny nuclei from their surrounding cellular gunk. Then, after more tests and experiments, he was startled to discover that they were made of a previously unknown substance. Neither a protein nor fat, the substance was acidic and had a high proportion of phosphorus not seen in any other organic materials. With no idea of what it was, Miescher named the substance "nuclein"—what we now call DNA.

Miescher published his findings in 1871 and went on to spend many years studying nuclein, isolating it from other cells and tissues. But its true nature remained a mystery. Although convinced that nuclein was critical to cell function, Miescher ultimately rejected the idea that it played a role in hereditary. Other scientists, however, were not so sure. In 1885, the Swiss anatomist Albert von Kolliker boldly claimed that nuclein must be the material basis of heredity. And in 1895, Edmund Beecher (E. B.) Wilson, author of the classic textbook *The Cell in Inheritance and Development*, agreed when he wrote:

> "...and thus we reach the remarkable conclusion that inheritance may, perhaps, be affected by the physical transmission of a particular chemical compound from parent to offspring."

And yet, poised on the brink of a world-changing discovery, science blinked—the world was simply not ready to embrace DNA as the biochemical "stuff" of heredity. Within a few years, nuclein was all-but-forgotten. Why did scientists give up on DNA for the next 50 years until 1944? Several factors played a role, but perhaps most important, DNA simply didn't seem up to the task. As Wilson noted in *The Cell* in 1925—a strong reversal from his praise in 1895—the "uniform" ingredients of nuclein seemed too uninspired compared to the "inexhaustible" variety of proteins. How could DNA possibly account for the incredible variety of life?

While the answer to that question would not unfold until the 1940s, Miescher's discovery had at least one major impact: It helped stimulate a wave of research that led to the *re*-discovery of a long-forgotten milestone. Not once, but *three* times.

Milestone #4 Born again: resurrection of a monastery priest and his science of heredity

Spring may be the season of renewal, but few events rivaled the rebirth in early 1900 when, after a 34-year hibernation, Gregor Mendel and his laws of heredity burst forth with a vengeance. Whether divine retribution for the lengthy oversight, or an inevitable outcome of the new scientific interest, in the early 1900 not one, but *three* scientists independently discovered the laws of heredity—and then realized that the laws had *already* been discovered decades earlier by a humble monastery priest.

Dutch botanist Hugo de Vries was the first to announce his discovery when his plant breeding experiments revealed the same 3 to 1 ratio seen by Mendel. Carl Correns, a German botanist, followed when his study of pea plants helped him rediscover the laws of segregation and independent assortment. And Austrian botanist Erich Tschermak published his discovery of segregation based on experiments in breeding peas begun in 1898. All three came upon Mendel's paper *after* their studies, while searching the literature. As Tschermak remarked, "I read to my great surprise that Mendel had already carried out such experiments much more extensive than mine, had noted the same regularities, and had already given the explanation for the 3:1 segregation ratio."

Although no serious disputes developed over who would take credit for the re-discovery, Tschermak later admitted to "a minor skirmish between myself and Correns at the Naturalists' meeting in Meran in 1903." But Tschermak added that all three botanists "were fully aware of the fact that [their] discovery of the laws of heredity in 1900 was far from the accomplishment it had been in Mendel's time since it was made considerably easier by the work that appeared in the interval."

As Mendel's laws of inheritance were reborn into the twentieth century, more and more scientists began to focus on those mysterious "units" of heredity. Although no one knew exactly what they were, by 1902 American scientist Walter Sutton and German scientist Theodor Boveri had figured out that they were located on chromosomes and that chromosomes were found in pairs in the cell. Finally, in 1909, Danish biologist Wilhelm Johannsen came up with a name for those units: genes.

Milestone #5 The first genetic disease: kissin' cousins, black pee, and a familiar ratio

The sight of black urine in a baby's diaper would alarm any parent, but to British physician Archibald Garrod, it posed an interesting problem in metabolism. Garrod was not being insensitive. The condition was called alkaptonuria, and though its most striking feature involved urine turning black after exposure to air, it is generally not serious and occurs in as few as one in one million people worldwide. When Garrod began studying alkaptonuria in the late 1890s, he realized the disease was not caused by a bacterial infection, as once thought, but some kind of "inborn metabolic disorder" (that is, a metabolic disorder that one is born with). But only after studying the records of children affected by the disease—whose parents in almost every case were first cousins—did he uncover the clue that would profoundly change our understanding of heredity, genes, *and* disease.

When Garrod first published the preliminary results of his study in 1899, he knew no more about genes or inheritance than anyone else, which explains why he overlooked one of his own key findings: When the number of children without alkaptonuria was compared to the number *with* the disease, a familiar ratio appeared: 3 to 1. That's right, the same ratio seen by Mendel in his second-generation pea plants (e.g., three purpled-flower plants to one white-flower plant), with its implications for the transmission of hereditary traits and the role of "dominant" and "recessive" particles (genes). In Garrod's study, the dominant trait was "normal urine" and the recessive trait "black urine," with the same ratio turning up in second-generation children: For every 3 children with normal urine, 1 child had black urine (alkaptonuria). Although Garrod didn't notice the ratio, it did not escape British naturalist William Bateson, who contacted Garrod upon hearing of the study. Garrod quickly agreed with Bateson that Mendel's laws suggested a new twist he hadn't considered: The disease appeared to be an *inherited* condition.

In a 1902 update of his results, Garrod put it all together—symptoms, underlying metabolic disorder, *and* the role of genes and inheritance. He proposed that alkaptonuria was determined by two hereditary "particles" (genes), one from each parent, and that the defective gene was recessive. Equally important, he drew on his biochemical background to propose how the defective "gene" actually *caused* the disorder: It must somehow produce a defective enzyme which, failing to perform its normal metabolic function, resulted in the black urine. With this interpretation, Garrod achieved a major milestone. He was proposing what genes actually *do*: They produce proteins, such as enzymes. And when a gene messed up—that is, was defective—it could produce a *defective* protein, which could lead to disease.

Although Garrod went on to describe several other metabolic disorders caused by defective genes and enzymes—including albinism, the failure to produce colored pigment in the skin, hair, and eyes—it would be another half-century before other scientists would finally prove him correct and appreciate his milestone discoveries. Today, Garrod is heralded as the first person to show the link between genes and disease. From his work developed the modern concepts of genetic screening, recessive inheritance, and the risks of interfamily marriage.

As for Bateson, perhaps inspired by Garrod's findings, he complained in a 1905 letter that this new branch of science lacked a good name. "Such a word is badly wanted," he wrote, "and if it were desirable to coin one, '*genetics*' might do."

***

In the early 1900s, despite a growing list of milestones, the new science was suffering an identity crisis, split as it was into two worlds. On the one hand, Mendel and his followers had established the laws of heredity but couldn't pinpoint what the physical "elements" were or where they acted. On the other hand, Flemming and others had discovered tantalizing physical landmarks in the cell, but no one knew how they related to heredity. In 1902, the worlds shifted closer together when American scientist Walter Sutton not only suggested that the "units" of hereditary were located on chromosomes, but that chromosomes were inherited as pairs (one from the mother and one from father) and that they "may constitute the physical basis of Mendelian laws of heredity." But it wasn't until 1910 that an American scientist—surprising himself as much as anyone else—joined these two worlds with one theory of inheritance.

Milestone #6 Like beads on a necklace: the link between genes and chromosomes

In 1905, Thomas Hunt Morgan, a biologist at Columbia University, was not only skeptical that chromosomes played a role in heredity, but was sarcastic of those at Columbia who supported the theory, complaining that the intellectual atmosphere was "saturated with chromosomic acid." For one thing, Morgan found the idea that chromosomes contained hereditary traits sounded too much like preformation, the once commonly believed myth that every egg contained an entire "preformed" person. But everything changed for Morgan around 1910, when he walked into the "fly room"—a room where he and his students had bred millions of fruit flies to study their genetic traits—and made a stunning discovery: One of his fruit flies had *white* eyes.

The rare white eyes were startling enough given that fruit flies normally had red eyes, but Morgan was even more surprised when he crossed the white-eyed male with a red-eyed female. The first finding was not so surprising. As expected, the first generation of flies all had red eyes, while the second generation showed the familiar 3 to 1 ratio (3 red-eyed flies for every 1 white-eyed fly). But what Morgan didn't expect, and what overturned the foundation of his understanding of heredity, was an entirely new clue: All of the white-eyed offspring were *male*.

This new twist—the idea that a trait could only inherited by one sex and not the other—had profound implications because of a discovery made several years earlier. In 1905, American biologists Nettie Stevens and E. B. Wilson had found that a person's gender was determined by two chromosomes, the so-called X and Y chromosomes. Females always had two X chromosomes, while males had one X and one Y. When Morgan saw that the white-eyed flies were always male, he realized that the gene for white eyes must somehow be linked to the male sex chromosome. That forced him to make a conceptual leap he'd been resisting for years: Genes must be a *part* of the chromosome.

A short time later, in 1911, one of Morgan's undergraduate students, Alfred Sturtevant, achieved a related milestone when he realized that the genes might actually be located along the chromosome in a *linear* fashion. Then, after staying up most of the night, Sturtevant produced the world's first gene map, placing five genes on a linear map and calculating the distances between them.

In 1915, Morgan and his students published a landmark book, *Mechanism of Mendelian Heredity*, that finally made the connection official: Those two previously separate worlds—Mendel's laws of heredity and the chromosomes and genes inside cells were one and the *same*. When Morgan won the 1933 Nobel Prize in Physiology or Medicine for the discovery, the presenter noted that the theory that genes were lined up on the chromosome "like beads in a necklace" initially seemed like "fantastic speculation" and "was greeted with justified skepticism." But when later studies proved him correct, Morgan's findings were now seen as "fundamental and decisive for the investigation and understanding of the hereditary diseases of man."

Milestone #7 A transformational truth: the rediscovery of DNA and its peculiar properties

By the late 1920s, many secrets of heredity had been revealed: The transmission of traits could be explained by Mendel's laws, Mendel's laws were linked to genes, and genes were linked to chromosomes. Didn't that pretty much cover everything?

Not even close. Heredity was still a mystery because of two major problems. First, the prevailing view was that genes were composed of proteins—*not* DNA. And second, no one had a clue how genes, whatever they were, *created* hereditary traits. The final mysteries began to unfold in 1928 when Frederick Griffith, a British microbiologist, was working on a completely different problem: creating a vaccine against pneumonia. He failed in his effort to make the vaccine but was wildly successful in revealing the next key clue.

Griffith was studying a type of pneumonia-causing bacteria when he discovered something curious. One form of the bacteria, the deadly S type, had a smooth outer capsule, while the other, the harmless R type, had rough outer surface. The S bacteria were lethal because they had a smooth capsule that enabled them to evade detection by the immune system. The R bacteria were harmless because, lacking the smooth outer capsule, they could be recognized and destroyed by the immune system. Then Griffith discovered something even stranger: If the deadly S bacteria were *killed* and mixed with harmless R bacteria, and both were injected into mice, the mice *still* died. After more experiments, Griffith realized that the previously harmless R bacteria had somehow "acquired" from the deadly S bacteria the ability to make the smooth protective capsule. To put it another way, even though the deadly S bacteria had been killed, something in them *transformed* the harmless R bacteria into the deadly S type.

What was that *something* and what did it have to do with heredity and genetics? Griffith would never know. In 1941, just a few years before the secret would be revealed, he was killed by a German bomb during an air raid on London.

***

When Griffith's paper describing the "transformation" of harmless bacteria into a deadly form was published in 1928, Oswald Avery, a scientist at the Rockefeller Institute for Medical Research in New York, initially refused to believe the results. Why should he? Avery had been studying the bacteria described by Griffith for the past 15 years, including the protective outer capsule, and the notion that one type could "transform" into another was an affront to his research. But when Griffith's results were confirmed, Avery became a believer and, by the mid-1930s, he and his associate Colin MacLeod had shown that the effect could be recreated in a Petri dish. Now the trick was to figure out exactly *what* was causing this transformation. By 1940, as Avery and MacLeod closed in on the answer, they were joined by a third researcher, Maclyn McCarty. But identifying the substance was no easy task. In 1943, as the team struggled to sort out the cell's microscopic mess of proteins, lipids, carbohydrates, nuclein, and other substances, Avery complained to his brother, "Try to find in that complex mixture the active principle! Some job—full of heart aches and heart breaks." Yet Avery couldn't resist adding an intriguing teaser, "But at last perhaps we have it."

Indeed they did. In February, 1944, Avery, MacLeod, and McCarty published a paper announcing that they had identified the "transforming principle" through a simple—well, not so simple—process of elimination. After testing everything they could find in that complex cellular mixture, only one substance transformed the R bacteria into the S form. It was nuclein, the same substance first identified by Friedrich Miescher almost 75 years earlier, and which they now called deoxyribonucleic acid, or DNA. Today, the classic paper is recognized for providing the first proof that DNA was *the* molecule of heredity. "Who would have guessed it?" Avery asked his brother.

In fact, few guessed *or* believed it because the finding flew in the face of common sense. How could DNA—considered by many scientists to be a "stupid" molecule and chemically "boring" compared to proteins—account for the seemingly infinite variety of hereditary traits? Yet even as many resisted the idea, others were intrigued. Perhaps a *closer* look at DNA would answer that other long-unanswered question: How did heredity *work*?

One possible clue to the mystery had been uncovered a few years earlier, in 1941, when American geneticists George Beadle and Edward Tatum proposed a theory that not only were genes *not* composed of proteins, but perhaps they actually *made* proteins. In fact, their research verified something Archibald Garrod had shown 40 years earlier with his work on the "black urine" disease, proposing that what genes *do* is make enzymes (a type of protein). Or, as it came to be popularly stated, "One gene, one protein."

But perhaps the most intriguing clue came in 1950. Scientists had known for years that DNA was composed of four "building block" molecules called bases: adenine, thymine, cytosine, and guanine. In fact, the repetitious ubiquity of those building blocks in DNA was a major reason why scientists thought DNA was too "stupid"—that is, too simplistic—to play a role in heredity. However, when Avery, MacLeod, and McCarty's paper came out in 1944, Erwin Chargaff, a biochemist at Columbia University, saw "the beginning of a grammar of biology…the text of a new language, or rather…where to look for it." Unafraid to take on that enigmatic book, Chargaff recalled, "I resolved to search for this text."

In 1949, after applying pioneering laboratory skills to analyze the components of DNA, Chargaff uncovered an odd clue. While different organisms had different amounts of those four bases, *all* living things seemed to share one similarity: The amount of adenine (A) and thymine (T) in their DNA was always about the same, as was the amount of cytosine (C) and guanine (G). The meaning of this curious one-to-one relationship—A&T and C&G—was unclear yet profound in one important way. It liberated DNA from the old "tetranucleotide hypothesis," which held that the four bases repeated monotonously, without variation in all species. The discovery of this one-to-one pairing suggested a potential for greater creativity. Perhaps DNA was not so dumb after all. And though Chargaff didn't recognized the importance of his finding, it led to the next milestone: the discovery of what heredity looked like—*and* how it worked.

Milestone #8 Like a child's toy: the secrets of DNA and heredity finally revealed

In 1895, Wilhelm Roentgen stunned the world and revolutionized medicine with the world's first X-ray—an eerie skeletal photograph of his wife's hand. Fifty-five years later, an X-ray image once again stunned the world and triggered a revolution in medicine. True, the X-ray image of DNA was far less dramatic than a human hand, looking more like lines rippling out from a pebble dropped in a pond than the skeletal foundation of heredity. But once that odd pattern had been translated into the DNA double helix—the famous winding staircase structure that scientist Max Delbrück once compared to "a child's toy that you could buy at the dime store"—the solution to an age-old mystery was at hand.

For James Watson, a graduate student at Cavendish Laboratory in England, the excitement began to build in May 1951, while he was attending a conference in Naples. He was listening to a talk by Maurice Wilkins, a New Zealand-born British molecular biologist at King's College in London, and was struck when Wilkins showed the audience an X-ray image of DNA. Although the pattern of fuzzy gray and black lines on the image was too crude to reveal the structure of DNA, much less its role in heredity, to Watson it provided tantalizing hints as to how the molecule might be arranged. Before long, it was proposed that DNA might be structured as a helix. But when Rosalind Franklin, another researcher at King's College, produced sharper images indicating that DNA could exist in two different forms, a debate broke out about whether DNA was really a helix after all.

By that time, Watson and fellow researcher Francis Crick had been working on the problem for about two years. Using evidence gathered by other scientists, they made cardboard cutouts of the various DNA components and built models of how the molecule might be structured. Then, in early 1953, as competition was building to be the first to "solve" the structure, Watson happened to be visiting King's College when Wilkins showed him an X-ray that Franklin had recently made—a striking image that showed clear helical features. Returning to Cavendish Laboratory with this new information, Watson and Crick reworked their models. A short time later, Crick had a flash of insight, and by the end of February 1953, all of the pieces fell into place: the DNA molecule *was* a double helix, a kind of endless spiral staircase. The phosphate molecules first identified by Friedrich Miescher in 1869 formed its two handrails, the so-called "backbone," while the base pairs described by Chargaff—A&T and C&G—joined to form its "steps."

When Crick and Watson published their findings in April 1953, their double helix model was a stunning achievement, not only because it described the structure of DNA, but because it suggested how DNA might *work*. For example, what exactly is a gene? The new model suggested a gene might be a specific *sequence* of base pairs within the double helix. And given that the DNA helix is *so* long—we now know there are about *3.1 billion* base pairs in every cell—there were more than enough base pairs to account for the genes needed to make the raw materials of living things, including hereditary traits. The model also suggested how those A&T and C&G sequences might actually make proteins: If the helix were to unwind and split at the point where two bases joined, the exposed single bases could act as a template that the cell could use to either make new proteins or—in preparation for dividing—a new chromosome.

Although Crick and Watson did not specify these details in their paper, they were well aware of the implications of their new model. "It has not escaped our notice that the [base] pairing we have postulated immediately suggests a possible copying mechanism for the genetic material." In a later paper, they added "...it therefore seems likely that the precise sequence of bases is the code which carries the genetic information." Interestingly, only a few months earlier Crick had been far less cautious in announcing the discovery, when he reportedly "winged" into a local pub and declared that he and Watson had discovered "the secret of life."

Milestone #9 The great recount: human beings have *how* many chromosomes?

By the time Crick and Watson had revealed the structural details of DNA in 1953, the world had known for years how many chromosomes are found in human cells. First described in 1882 by Walther Flemming, chromosomes are the tiny paired structures that DNA twists, coils, and wraps itself into. During the next few decades, *everyone* had seen chromosomes. And though difficult to see and count given technological limits of the time, by the early 1920s geneticist Thomas Painter was confident enough to boldly proclaim the number that would be universally accepted around the world: 48.

Wait—*what*?

In fact, it was not until 30 years later, in 1955, that Indonesian-born scientist Joe-Hin Tjio discovered that human cells actually have *46* chromosomes (arranged in 23 pairs). The finding—announced to a somewhat red-faced scientific community in 1956—was made possible by a technique that caused chromosomes to spread apart on a microscope slide, making them easier to count. In addition to nailing down the true number, the advance helped establish the role of cell genetics in medicine and led to subsequent discoveries that linked chromosomal abnormalities with specific diseases.

Milestone #10 Breaking the code: from letters and words to a literature of life

Crick and Watson may have discovered the secret to life in 1953, but one final mystery remained: How do cells use those base pair "steps" inside the DNA helix to actually *build* proteins? By the late 1950s, scientists had discovered some of the machinery involved—including how RNA molecules help "build" proteins by ferrying raw materials around inside the cell—but it was not until two years later that they finally cracked the genetic code and determined the "language" by which DNA makes proteins.

In August 1961, American biochemist Marshall Nirenberg and his associate J. Heinrich Matthaei announced the discovery of the first "word" in the language of DNA. It consisted of only three letters, with each letter being one of the four bases arranged in a specific order and, in turn, coding for other molecules used to build proteins. And thus the genetic code was broken. By 1966, Nirenberg had identified more than 60 of the so-called "codons," each representing a unique three-letter word. Each three-letter word was then used to build one of 20 "sentences," the 20 major amino acids that form the building blocks of protein. And from those protein sentences sprang the story of life, the countless biological substances found in all living things—from enzymes and hormones, to tissues and organs, to the hereditary traits that make us each unique.

By the end of 1961, as news spread across the world that the "code of life" had been broken, public reaction spanned a predictable range of extremes. An article in the *Chicago Sun-Times* optimistically offered that with the new information, "science may deal with the aberrations of DNA arrangements that produce cancer, aging, and other weaknesses of the flesh." Meanwhile, one Nobel Laureate in chemistry warned that the knowledge could be used for "creating new diseases [and] controlling minds."

By then, of course, Nierenberg had heard it all. In 1962, he wrote to Francis Crick, noting dryly that the press "has been saying [my] work may result in (1) the cure of cancer and allied diseases (2) the cause of cancer and the end of mankind, and (3) a better knowledge of the molecular structure of God." But Nirenberg shrugged it off with good humor. "Well, it's all in a day's work."

***

And so finally, after thousands of years of speculation, misconception, and myth, the secret of heredity, genetics, and DNA had been revealed. In many ways, the breakthrough was more than anyone could have imagined or bargained for. With the "blueprint" laid out before us, the molecular details of our wiring exposed from the inside-out, humanity collectively changed its entire way of thinking—and worrying—about every aspect of ourselves and our lives. Hidden in the microscopic coils of our DNA was the explanation for everything: the admirable and not-so-admirable traits of ourselves and families, the origins of health and disease, perhaps even the structural basis of good, evil, God, and the cosmos.

Well, not quite. As we now know, DNA has turned out to be considerably more complicated than that—except, perhaps, for the worrying.

But what's unquestionably true and what forever changed our way of thinking was the discovery that the genetic code is a *universal* language. And as we sort out the details of how genes influence hereditary traits, health, and disease, one underlying truth is that the same genetic machinery underlies *all* living things, unifying life in a way whose significance we may not comprehend for years to come.

Fifty years later: more milestones and more mysteries

Fifty years after the genetic code was broken, a funny thing happened on the way to one of medicine's top ten breakthroughs: It never quite *arrived*. Since the early 1960s, new milestones continue to appear like a perpetually breaking wave, each discovery reshaping the shoreline of an ongoing revolution. Consider just a few recent milestones:

- **1969**: Isolation of the first individual gene (a segment of bacterial DNA that assists in sugar metabolism)
- **1973**: Birth of genetic engineering (a segment of frog DNA is inserted into a bacterial cell, where it is replicated)
- **1984**: Birth of genetic fingerprinting (use of DNA sequences for identification)
- **1986**: First approved genetically engineered vaccine (hepatitis B)
- **1990**: First use of gene therapy
- **1995**: DNA of a single-cell organism decoded (influenza bacterium)
- **1998**: DNA of a multicellular organism decoded (roundworm)
- **2000**: DNA of a human decoded (a "working draft" that was subsequently completed in 2003)

When researchers at the Human Genome Project announced the decoding of the human genome in 2000, it helped initiate a new age of genetics that is now laying the groundwork for a widespread revolution in biology and medicine. Today, work from this project and other researchers around the globe has opened new insights into human ancestry and evolution, uncovering new links between genes and disease, and leading to countless other advances that are now revolutionizing medical diagnosis and treatment.

Gregor Mendel, who in 1865 dared not even speculate on what the "elements" of hereditary might be, would surely be amazed: We now know that humans have about 25,000 genes—far fewer than the *80,000 to 140,000* that some once believed—and comparable to some far simpler life forms, including the common laboratory mouse (25,000 genes), mustard weed (25,000 genes), and roundworm (19,000 genes). How could a mouse or weed have as many genes as a human being? Scientists believe that the complexity of an organism may arise not only from the number of genes, but in the complex way different *parts* of genes may interact. Another recent surprise is that genes comprise only *2%* of the human genome, with the remaining content probably playing structural and regulatory roles. And, of course, with new findings come new mysteries: We still don't know what *50%* of human genes actually *do*, and despite the celebrated diversity of humans, how is it that the DNA of all people is *99.9% identical?*

It is now clear that the answer to such questions extends beyond the DNA we inherit to the larger world around us. Humans have long suspected that heredity alone cannot account for our unique traits or susceptibility to disease, and new discoveries are now shedding light on one of the greatest mysteries of all: How do genes *and* environment interact to make us who we are?

The secret life of snips and the promise of genetic testing

For most of us, it's difficult to imagine 3.1 billion pairs of *anything*, much less the chemical bases that comprise the DNA found in any given cell. So try instead to imagine 3.1 billion pairs of shoes extending into outer space. Now *double* that number so you're picturing 6 billion *individual* shoes stretching into the cosmos. Six billion is the number of individual bases in the human genome, which scientists call "nucleotides." As inconsequential as any one nucleotide may seem, a change to just *one* nucleotide—called a single nucleotide polymorphism, or SNP (pronounced "snip")—can significantly impact human traits *and* disease. If you doubt that one SNP among billions could have much importance, consider that sickle cell anemia arises from just one SNP, and that most forms of cystic fibrosis arise from the loss of three nucleotides. As of 2008, researchers had identified about 1.4 million SNPs in the human genome. What causes these miniscule alterations to our DNA? Key suspects include environmental toxins, viruses, radiation, and errors in DNA copying.

The good news is that current efforts to identify SNPs are not only helping uncover the causes of disease, but creating chromosome "landmarks" with a wide range of applications. In 2005, researchers completed the first phase of a project in which they analyzed the DNA of people throughout the world and constructed a "map" of such landmarks based on 500,000 or more SNPs. This information is now revealing links between tiny genetic variations and specific diseases, which in turn is leading to new approaches to diagnosis (e.g., genetic testing) and treatment. For example, in the growing field of pharmacogenomics, doctors can use such information to make personalized treatment decisions based on a person's genetic makeup. Recent examples include genetic tests that can identify forms of breast cancer that are susceptible to certain drugs or that can identify patients who may be susceptible to dangerous side effects when receiving the clot-preventing drug warfarin (Coumadin).

SNP findings are also providing insights into age-old questions such as how much we are influenced by genes and environment. In fact, it's becoming increasingly clear that many common diseases—including diabetes, cancer, and heart disease—are probably caused by a complex interaction between *both*. In the relatively new field of epigenetics, scientists are looking at the netherworld where the two may meet—that is, how "external" factors such as exposure to environmental toxins may affect a person's SNPs and thus susceptibility to disease.

Unfortunately, researchers are also learning that sorting out the role of SNPs and disease is a very complex business. The good news about the HapMap project is that researchers have already uncovered genetic variants associated with the risk of more than *40 diseases*, including type 2 diabetes, Crohn's disease, rheumatoid arthritis, elevated blood cholesterol, and multiple sclerosis. The bad news is that many diseases and traits are associated with so *many* SNPs that the meaning of any one variation is hard to gauge. According to one recent estimate, 80% of the variation in height in the population could theoretically be influenced by as many as *93,000* SNPs. As David B. Goldstein wrote in a 2009 issue of the *New England Journal of Medicine (NEJM)*, if the risk for a disease involves many SNPs, with each contributing just a small effect, "then no guidance would be provided: In pointing at everything, genetics would point at nothing."

While many of us are curious to try the latest genetic tests and learn about our risk for disease, Peter Kraft and David J. Hunter cautioned in the same issue of *NEJM* that, "We are still too early in the cycle of discovery for most tests... to provide stable estimates of genetic risk for many diseases." However, they add that rapid progress is being made and that "the situation may be very different in just 2 or 3 years." But as better tests become available, "Appropriate guidelines are urgently needed to help physicians advise patients...as to how to interpret, and when to act on, the results."

"We'll figure this out": the promise of gene therapy

To some people, 1990 was *the* breakthrough year for genetics and medicine. In that year, W. French Anderson and his colleagues at the National Institutes of Health performed the first successful gene therapy in a four-year-old girl who had an immune deficiency disease caused by a defective gene that normally produces an enzyme called ADA. Her treatment involved an infusion of white blood cells with a corrected version of the ADA gene. But though the results were promising and spurred hundreds of similar clinical trials, a decade later it was clear that few gene therapy trials were actually *working*. The field suffered another setback in 1999, when 18-year-old Jesse Gelsinger received gene therapy for a non-life-threatening condition. Within days, the therapy itself killed Gelsinger, and the promise of gene therapy seemed to come to a crashing halt. But as one shaken up doctor said at Gelsinger's bedside at the time of his death, "Good bye, Jesse...We'll figure this out."

Ten years later, scientists are beginning to figure it out. While the challenges of gene therapy are many—two major issues are how to safely deliver repaired genes into the body and how to ensure that the patient's body accepts and uses them—many believe that the technique will soon be used to treat many genetic diseases, including blood disorders, muscular dystrophy, and neurodegenerative disorders. Recent progress includes modest successes in treating hereditary blindness, HIV, and rheumatoid arthritis. And in 2009, researchers reported on a follow-up study in which 8 out of 10 patients who received gene therapy for the defective ADA gene had "excellent and persistent" responses. As Donald B. Kohn and Fabio Candotti wrote in a 2009 editorial in the *NEJM*, "The prospects for continuing advancement of gene therapy to wider applications remain strong" and may soon "fulfill the promise that gene therapy made two decades ago."

In other words, the breakthrough has arrived and *continues* to arrive. With the double helix unwinding in so many directions—spinning off findings that impact so many areas of science, society, and medicine—we can be patient. Like Hippocrates' indulgence of the woman who gazed too long at the portrait of an Ethiopian on her wall, like Mendel's years of counting thousands of pea plant traits, and like the milestone work of countless researchers over the past 150 years—we can be patient. It's a long road, but we've come a long way.

***

Up until the early 1800s, many scientists believed—like Hippocrates 2,200 years before them—that maternal impression was a reasonable explanation for how a mother might pass some traits on to her unborn child. After all, perhaps the shock of what a pregnant mother witnessed was somehow transferred to her fetus through small connections in the nervous system. But by the early 1900s, with advances in anatomy, physiology, and genetics providing other explanations, the theory of maternal impression was abandoned by most physicians.

Most, but not all...

- *In the early 1900s, the son of a pregnant mother was violently knocked down by a cart. He was rushed to the hospital, where the mother could not restrain herself from a fearful glance as the physician stitched up her son's bloodied scalp. Seven months later she gave birth to a baby girl with a curious trait: There was a bald area on her scalp in exactly the same region and size as where her brother had been wounded.*

This story, along with 50 other reports of maternal impression, appeared in an article published in a 1992 issue of the Journal of Scientific Exploration. *The author, Ian Stevenson, a physician at the University of Virginia School of Medicine, did not mention genetics, made no attempt to provide a scientific explanation, and noted, "I do not doubt that many women are frightened during a pregnancy without it having an ill effect on their babies." Nevertheless, based on his analysis, Stevenson concluded, "In rare instances maternal impressions may indeed affect gestating babies and cause birth defects."*

In the brave new world of genes, nucleotides, and SNPs, it's easy to dismiss such mysteries as playing no role in the inheritance of physical traits—no more than, say, DNA was thought to have for 75 years after its *discovery.*

Medicines for the Mind: THE DISCOVERY OF DRUGS FOR MADNESS, SADNESS, AND FEAR

9

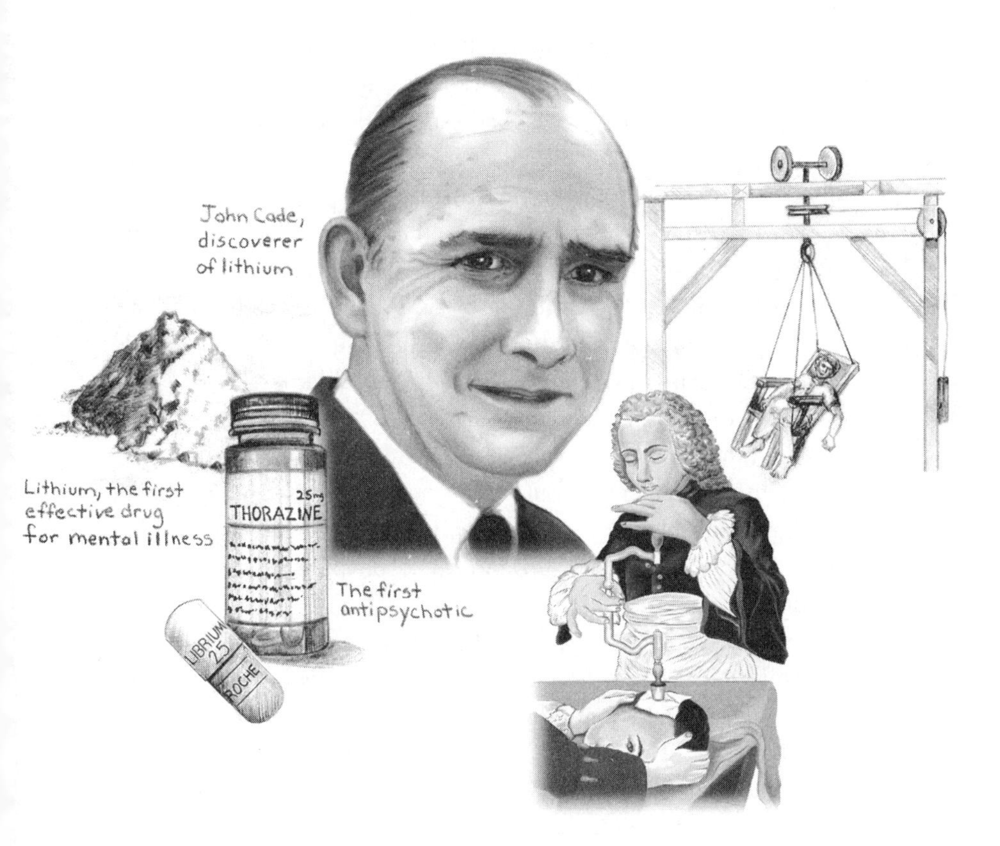

One cold evening in February, 2008, a 39-year-old man dressed in a dark hat, trench coat, and sneakers ducked out of the freezing rain and into a suite of mental health offices several blocks east of Central Park in New York City. Toting two black suitcases, he climbed a short flight of steps, entered a waiting room, and—directed by the voice of God—prepared to rob psychiatrist Dr. Kent Shinbach. But the man was apparently in no hurry; told that Dr. Shinbach was busy, he put his luggage aside, sat down, and for the next 30 minutes made small talk with another patient.

Then something went terribly wrong. For no clear reason, the man suddenly got up and entered the nearby office of psychologist Kathryn Faughey. Armed with two knives and a meat cleaver, he erupted in a frenzied rage, viciously slashing and stabbing the 56-year-old therapist in the head, face, and chest, splattering blood across the walls and furniture. Hearing the screams, Dr. Shinbach rushed out of his office to help, only to find Dr. Faughey's lifeless body on the blood-soaked carpet. Before the 70-year-old psychiatrist could escape, the man began attacking him, stabbing him in the face, head, and hands. The man finally pinned Dr. Shinbach against a wall with a chair, stole $90 from his wallet, and fled. Dr. Shinbach survived, but Dr. Faughey—who was described as a "good person" who "changed and saved people's lives"—was pronounced dead at the scene.

It wasn't until several days later that police arrested and charged David Tarloff with the murder and the bizarre details began to unfold. "Dad, they say I killed some lady," Tarloff said in a phone call to his father at the time of his arrest. "What are they talking about?" As Tarloff's dazed words suggested, what everyone would soon be talking about was not only the evidence pointing to his guilt, but his full-blown insanity. As *The New York Times* reported in subsequent weeks, Tarloff lived in nearby Queens and had been diagnosed with schizophrenia 17 years earlier, at the age of 22. The diagnosis was made by Dr. Shinbach, who no longer even remembered Tarloff or the diagnosis. According to Tarloff, his only motive the night of the murder had been to rob Dr. Shinbach. Dr. Faughey had somehow crossed his path, resulting in a random and tragically meaningless death. Why the sudden need for cash? Tarloff explained that he wanted to free his mother from a nursing home so they could "leave the country."

Which is where things really get interesting.

Tarloff's mother *was* in a nearby nursing home, but Tarloff's interest had long-since escalated from healthy concern to pathologic obsession, with "harassing" visits, frequent daily phone calls, and, most recently, when he was found lying in bed with her in the ICU. According to his father, Tarloff had a long history of mental health problems. In addition to bouts of depression, anxiety, and mania, he'd displayed such obsessive-compulsive behaviors as taking 15 or 20 showers a day and calling his father 20 or more times after an argument to say "I'm sorry" in just the right way. But perhaps most striking were the recent symptoms suggesting that Tarloff had completely lost touch with reality. Apart from the auditory hallucinations in which God encouraged him to rob Dr. Shinbach, Tarloff's paranoia and mental confusion continued after his arrest, when he blurted out in a courtroom, "If a fireman comes in, the police come in here, the mayor calls, anyone sends a messenger, they are lying. The police are trying to kill me."

***

Who is David Tarloff, and how did he get this way? By most accounts, Tarloff had been a well-adjusted youth with a relatively normal childhood. One Queens neighbor reported that Tarloff had always been "the ladies man, tall and thin…with tight jeans and always good looking." His father agreed that Tarloff had been "handsome, smart, and happy" while growing up but that something changed as he entered young adulthood and went to college. When he returned, his father recalled, Tarloff was moody, depressed, and silent. He "saw things" and believed people were "against him." Unable to hold a job, he dropped out of two more colleges. After his diagnosis of schizophrenia, and over the next 17 years, Tarloff had been committed to mental hospitals more than a dozen times and prescribed numerous antipsychotic drugs. Yet despite some petty shoplifting and occasionally bothering people for money, neighbors viewed him with more pity than fear. One local store clerk described Tarloff as "a sad figure whose stomach often hung out, with his pants cuffs dragging and his fly unzipped."

But in the year leading up to the murder, Tarloff's mental health seemed to worsen. Eight months earlier, he had threatened to kill everyone at the nursing home where his mother was kept. Two months after that, police were called to his home by reports that he was behaving violently. And two weeks before the murder, he attacked a nursing home security guard. Tarloff's mental state was clearly deteriorating, and it continued to decline after the murder. A year later, while awaiting trial in a locked institution, he claimed that he was the Messiah and that DNA tests would prove him to be the son of a nearby inmate, who himself believed he was God. Doctors despaired that Tarloff "may never be sane enough to understand the charges against him."

Mental illness today: the most disorderly of disorders

Madness, insanity, psychosis, hallucinations, paranoia, delusions, incoherency, mania, depression, anxiety, obsessions, compulsions, phobias...

Mental illness goes by many names, and it seems that David Tarloff had them all. But Tarloff was lucky in one way: After centuries of confused and conflicting attempts to understand mental disorders, physicians are now better than ever at sorting through the madness. But though most psychiatrists would agree with the specific diagnosis given to Tarloff—acute paranoid schizophrenia—such precision cannot mask two unsettling facts: The symptoms experienced by Tarloff also occur in many other mental disorders, and his treatment failure led to the brutal murder of an innocent person. *Both* facts highlight a nasty little secret about mental illness: We still don't know exactly what it is, what causes it, how to diagnose and classify it, or the best way to treat it.

Oops.

That's not to discount the many advances that have been made over the centuries, which range from the realization in ancient times that mental illness is caused by natural factors rather than evil spirits, to the milestone insight in the late-eighteenth century that mentally ill patients fared better when treated with kindness rather than cruelty, to one of the top ten breakthroughs in the history of medicine: The mid-twentieth century discovery of the first effective drugs for madness, sadness, and fear—more commonly known as schizophrenia and manic-depression (madness), depression (sadness), and anxiety (fear).

Yet despite these advances, mental illness remains a unique and vexing challenge compared to most other human ailments. On the one hand, they can be as disabling as any "physical" disease; in addition to often lasting a lifetime, destroying the lives of individuals and families and crippling careers, they can also be fatal, as in cases of suicide. On the other hand, while most diseases have a generally known cause and leave a trail of evidence—think of infections, cancer, or the damaged blood vessels underlying heart disease—mental disorders typically leave no physical trace. Lacking objective markers and a clear link between cause and effect, they cannot be diagnosed with laboratory tests and present no clear divisions between one condition and another. All of these factors stymie the search for treatments, which—no surprise—tend to work better when you actually know what you're treating.

Because of such challenges, the "Bible" for diagnosing mental disorders in the United States—the American Psychiatric Association's *Diagnostic and Statistical Manual of Mental Disorders (DSM-IV)*—provides guidelines based mainly on "descriptive" symptoms. But as seen with David Tarloff and many patients, descriptive symptoms can be subjective, imprecise, and not exclusive to any one disorder. Even the DSM-IV—all 943 pages of it—states that "It must be admitted that no definition adequately specifies precise boundaries for the concept of a mental disorder."

At least there *is* agreement on a general definition of mental illness and how debilitating it can be. According to the National Alliance on Mental Illness (NAMI), mental disorders are "medical conditions that disrupt a person's thinking, feeling, mood, ability to relate to others, and daily functioning." In addition, they often "diminish a person's ability to cope with the ordinary demands of life," affect people of all ages, races, religion, or income, and are "not caused by personal weakness."

Recent studies have also provided eye-opening statistics as to how common and serious mental disorders are. The World Health Organization (WHO) recently found that about 450 million people worldwide suffer from various mental illnesses and that nearly 900,000 people commit suicide each year. Just as alarming, a 2008 WHO report found that based on a measure called "burden of disease"—defined as premature death combined with years lived with a disability—depression now ranks as the fourth most serious disease in the world. By 2030, it will be the world's *second* most serious burden of disease, after HIV/AIDS.

But perhaps the biggest barrier to understanding and treating mental illness is the sheer number of ways the human mind can go awry: The *DSM-IV* divides mental illness into as many as 2,665 categories. Although such fragmentation of human suffering can be helpful to researchers, a 2001 WHO report found that the most serious mental illnesses fall into just four categories that rank among the world's top ten causes of disability: Schizophrenia, Bipolar Disorder (manic-depression), Depression, and Anxiety. The identification of these four disorders is intriguing because, as it turns out, one of the ten greatest breakthroughs in medicine was the discovery of drugs for these *same* conditions—antipsychotic, antimanic, antidepressant, and anti-anxiety drugs. What's more, physicians have been struggling to understand these disorders for thousands of years.

The many faces of madness: early attempts to understand mental illness

> "He huddled up in his clothes and lay not knowing where he was. His wife put her hand on him and said, 'No fever in your chest. It is sadness of the heart.'"
>
> —*Ancient Egyptian papyri, c. 1550 BC*

Descriptions of the four major types of mental illness date back to the dawn of civilization. In addition to this account of depression, reports of schizophrenia-like madness can be found in many ancient documents, including Hindu Vedas from 1400 BC that describe individuals under the influence of "devils" who were nude, filthy, confused, and lacking self-control. References to manic-depression—periods of hyper-excited and grandiose behavior alternating with depression—can be found as early as the second century AD in the writings of Roman physician Soranus of Ephedrus. And in the fourth century BC, Aristotle described the debilitating effects of anxiety, linking it to such physical symptoms as heart palpitations, paleness, diarrhea, and trembling.

The most durable theory of mental illness, dating back to the fourth century BC and influential up until the 1700s, was Hippocrates' humoral theory, which stated that mental illness could occur when the body's four "humors"—phlegm, yellow bile, black bile, and blood—became imbalanced. Thus, excess phlegm could lead to insanity; excess yellow bile could cause mania or rage; and excess black bile could result in depression. While Hippocrates was also the first to classify paranoia, psychosis, and phobias, later physicians devised their own categories. For example, in 1222 AD Indian physician Najabuddin Unhammad listed seven major types of mental illness, including not only madness and paranoia, but "delusion of love."

But perhaps the most breathtakingly simple classification was seen in the Middle Ages and illustrated by the case of Emma de Beston. Emma lived in England at a time when mental illness was divided into just two categories: Idiots and Lunatics. The division wasn't entirely crazy. Based on legal custom of the day, idiots were those born mentally incompetent and whose inherited profits went to the king; lunatics were those who had lost their wits during their lifetimes and whose profits stayed with the family. According to well-documented legal records of the time, on May 1, 1378, Emma's mind was caught "by the snares of evil spirits," as she suddenly began giving away a large part of her possessions. In 1383, at her family's request, Emma was brought before an inquisition to have her mental state assessed. Her responses to their questions were revealing: She knew how many days were in the week but could not name them; how many men she had married (three) but could name only two; and she could not name her son. Based on her sudden mental deterioration, Emma was judged a lunatic—presumably to the satisfaction *and* profit of her family.

By the sixteenth and seventeenth centuries, with the growing influence of the scientific revolution, physicians began to take a harder look at mental illness. In 1602, Swiss physician Felix Platter published the first medical textbook to discuss mental disorders, noting that they could be explained by both Greek humoral theory and the work of the devil. Another key milestone came in 1621, when Robert Burton, a vicar and librarian in Oxford, England, published *The Anatomy of Melancholy*, a comprehensive text on depression that rejected supernatural causes and emphasized a humane view. Depression, wrote Burton, "is a disease so grievous, so common, I know not [how to] spend my time better than to prescribe how to prevent and cure a malady that so often crucifies the body and the mind." Burton also provided vivid descriptions of depression, including: "It is a chronic disease having for its ordinary companions fear and sadness without any apparent occasion… [It makes one] dull, heavy, lazy, restless, unapt to go about any business."

Although physicians struggled for the next two centuries to understand madness, one early milestone came in 1810, when English physician John Haslam published the first book to provide a clear description of a patient with schizophrenia. The patient, James Tilly Matthews, believed an "internal machine" was controlling his life and torturing him—an interesting delusion given that Matthews lived at the dawn of the Industrial Revolution. Haslam also summarized the confusion of many physicians when he wrote that madness is "a complex term for all its forms and varieties. To discover an infallible definition… I believe will be found impossible." But physicians did not give up, and in 1838 French psychiatrist Jean Etienne Dominique Esquirol published the first modern treatise on mental disorders, in which he introduced the term "hallucinations" and devised a classification that included paranoia, obsessive compulsive disease, and mania.

In the meantime, by the 1800s, the term "anxiety" was beginning to appear in the medical literature with increasing frequency. Until then, anxiety had often been viewed as a symptom of melancholy, madness, or physical illness. In fact, confusion over where anxiety fit into the spectrum of mental illness underwent many shifts over the next two centuries, from Sigmund Freud's 1894 theory that "anxiety neurosis" was caused by a "deflection of sexual excitement," to the twentieth-century realization that the "shell shock" suffered by soldiers in wartime was a serious mental disorder related to anxiety. Although the APA did not include "anxiety" in its manuals until 1942, today the DSM-IV lists it as a major disorder, with subcategories that include panic disorder, OCD, posttraumatic stress disorder (PTSD), social phobia, and various specific phobias.

Although the profession of psychiatry was "born" in the late 1700s, madness remained a vexing problem throughout most of the nineteenth century. The problem was that the symptoms of insanity could be so varied—from angry fits of violence to the frozen postures and stone silence of catatonia; from bizarre delusions and hallucinations to manic tirades of hyper-talkativeness. But in the late 1890s, German psychiatrist Emil Kraepelin made a landmark discovery. After studying thousands of psychotic patients and documenting how their illness progressed over time, Kraepelin was able to sort "madness" into two major categories: 1) Manic-Depression, in which patients suffered periods of mania and depression but did *not* worsen over time; and 2) Schizophrenia, in which patients not only had hallucinations, delusions, and disordered thinking, but often developed their symptoms in young adulthood and *did* worsen over time. Although Kraepelin called the second category "dementia praecox," the term schizophrenia was later adopted to reflect the "schisms" in a patient's thoughts, emotions, and behavior.

Kraepelin's discovery of the two new categories of madness was a major milestone that remains influential to this day and is reflected in the DSM-IV. In fact, David Tarloff is a textbook case of schizophrenia because his symptoms (hallucinations, paranoia, delusions, and incoherent speech) started in young adulthood and worsened over time. Kraepelin's insight not only helped clear the foggy boundaries between two major mental disorders, but set the stage for the discovery of drugs to treat mental illness—a remarkable achievement given the alarming array of treatments used in the previous 2,500 years...

Bloodletting, purges, and beatings: early attempts to "master" madness

> "Madmen are strong and robust... They can break cords and chains, break down doors or walls, easily overthrow many endeavoring to hold them. [They] are sooner and more certainly cured by punishments than medicines."
>
> —*Thomas Willis, 1684*

Treatments for mental illness have a long history of better serving the delusions of those administering them than the unfortunate patients receiving them. For example, while Hippocrates' treatments undoubtedly made sense to those who believed in humoral theory, his prescriptions for eliminating excess bile and phlegm—bleeding, vomiting, and strong laxatives—were probably of little comfort to his patients. And while some ancient treatments involved gentler regimens, such as proper diet, music, and exercise, others were literally more frightening: In the thirteenth century AD, Indian physician Najabuddin Unhammad's prescriptions for mental illness included scaring patients into sanity through the use of snakes, lions, elephants, and "men dressed as bandits."

The Middle Ages also saw the rise of institutions to care for the mentally ill, but with decidedly mixed results. On the plus side, the religion of Islam taught that society should provide kindly care for the insane, and followers built hospitals and special sections for the mentally ill, including facilities in Baghdad (750 AD) and Cairo (873 AD). In contrast, perhaps the most famous and most notorious asylum in Europe was

London's Bethlem hospital, which began admitting insane patients around 1400. Over the next century, Bethlem became dominated by patients with severe mental illness, leading to its reputation as a "madhouse" and the popular term based on its name, "bedlam." And that's when the trouble *really* began.

Throughout the 1600s and 1700s, imprisonment and mistreatment of the insane in European asylums like Bethlem occurred with alarming frequency. Out of stigmatization and fear, society began to view mentally ill patients as incurable wild beasts who had to be restrained with chains and tamed with regular beatings and cruel treatments. "Madmen," wrote English physician Thomas Willis in a 1684 book, "are almost never tired... They bear cold, heat, fasting, strokes, and wounds without sensible hurt." While Willis was probably referring to extreme cases of schizophrenia and mania, the public was intrigued—and amused—to watch from a distance: At one time, up to 100,000 people visited Bethlem yearly, happy to pay the one penny admission fee to view the madmen and their "clamorous ravings, furious gusts of outrageous action, and amazing exertion of muscular force."

Asylum directors, meanwhile, focused on "treatments" to keep their patients under control. While Willis contended that "bloodletting, vomits, or very strong purges are most often convenient," others suggested that the best way to gain "complete mastery" over madmen was near-drowning. To that end, creative therapies included hidden trapdoors in corridors to drop unsuspecting lunatics into a "bath of surprise" and coffins with holes drilled into the cover, in which patients were enclosed and lowered into water. But perhaps the cruelest "therapy" of all was the rotating/swinging chair, as described by Joseph Mason Cox in 1806. Patients were strapped into a chair that hung from several chains and that an operator could simultaneously swing and/or revolve "with extraordinary precision." Cox wrote that with a series of maneuvers—increasing the swing's velocity, quick reversals, pauses, and sudden stopping—a skilled operator could trigger "an instant discharge of the stomach, bowels, and bladder, in quick succession."

While the mistreatment of mentally ill patients continued throughout most of the eighteenth century, a key milestone occurred in the late-1700s when French physician Philippe Pinel began a movement he called the "moral treatment of insanity." In 1793, Pinel had become director of a men's insane asylum at Bicetre. Within a year, he had developed a new philosophy and approach to treating mental illness based on carefully observing and listening to patients, recording the history of their illness, and treating them "in a psychologically sensitive way." In his famous 1794 *Memoir on Madness*, Pinel wrote that "One of the most fundamental principles of conduct one must adopt toward the insane is an intelligent mixture of affability and firmness." Pinel also strongly opposed physical restraints unless absolutely necessary. In 1797, following a similar move by an associate at Bicetre, Pinel famously unshackled patients at Salpetriere, a public hospital for women. Today, Pinel is considered to be the father of psychiatry in France.

Unfortunately, although the moral treatment promoted by Pinel and others was influential throughout the 1800s, the model eventually failed as increasing numbers of patients became "warehoused" in large, crowded institutions. By the end of the nineteenth century, other trends had begun to dominate the field of mental illness, including a growing emphasis on the anatomy and physiology of the nervous system and new psychological approaches developed by Freud and his followers. But though Freud's talk-based treatments were influential in the United States and were a vital precursor to modern psychotherapy, they eventually fell out of favor due to their ineffectiveness in serious mental illness and lack of a biological foundation.

And so by the early 1900s, after centuries of dismal failure, the world was ready for a new approach to treating mental illness. The first milestone finally arrived in the form of several "medical" treatments that ranged from the frightening to the bizarre. But at least they worked—sort of.

Milestone #1 Sickness, seizure, surgery, and shock: the first medical treatments for mental illness

Sickness. Insanity can have many causes, but probably one of the most humbling is syphilis, a sexually transmitted disease. Today syphilis is easily treated and cured with penicillin, but in the early 1900s it often progressed to its final stages, when it can attack the brain and nerves and cause, among other symptoms, insanity. In 1917, Austrian psychiatrist Julius Wagner-Jauregg decided to investigate a possible treatment for this one cause of mental illness based on an idea he'd been thinking about for 30 years: He would attempt to cure one devastating disease, syphilis, with another: malaria

The idea wasn't without precedent. Malaria produces fever, and physicians had long known that, for unknown reasons, mental illness sometimes improved after a severe fever. And so in 1917, Wagner-Jauregg injected nine patients with a mild, treatable form of malaria. The patients soon developed a fever, followed by a side effect that Wagner-Jauregg called "gratifying beyond expectation." The mental disturbances in all nine patients improved, with three being "cured." Malaria treatment was subsequently tested throughout the world, with physicians reporting cure rates as high as 50%, and Wagner-Jauregg won the Nobel Prize in 1927 for his discovery. Although his treatment only addressed one type of insanity caused by an infection that is easily prevented today, it showed for the first time that mental illness *could* be medically treated.

Seizure. In 1927, Polish physician Manfred Sakel discovered that too much of a good thing—insulin—can be bad *and* good. Normally, the body needs insulin to metabolize glucose and thus prevent diabetes. Sakel had discovered that when a morphine-addicted woman was accidentally given an overdose of insulin and fell into a coma, she later woke up in an improved mental state. Intrigued, he wondered if a similar "mistake" might help other patients with mental illness. Sure enough, when he gave insulin overdoses to patients with schizophrenia, they experienced coma and seizures, but also recovered with improved mental functioning. Sakel reported his technique in 1933, and it was soon being hailed as the first effective medical treatment for schizophrenia. Within a decade, "insulin shock" therapy had spread throughout the world, with reports that more than 60% of patients were helped by the treatment.

While Sakel was experimenting with insulin, others were pursuing a different but related idea. Physicians had observed that epilepsy was rare in patients with schizophrenia, but in those who did have epilepsy, their mental symptoms often improved after a seizure. This begged the question: Could schizophrenia be treated by *purposely* triggering seizures? In 1935, Hungarian physician Ladislaus von Meduna, who was experienced in both epilepsy and schizophrenia, induced seizures in 26 patients with schizophrenia by injecting them with a drug called metrazol (cardiazol). Although the effect was unnerving—patients experienced quick and violent convulsions—the benefit was impressive, with 10 of the 26 patients recovering. Subsequent studies found that up to 50% of patients with schizophrenia could be discharged after treatment and that some had "dramatic cures." When Meduna reported his results in 1937, insulin therapy was well-known, leaving physicians with a choice: Metrazol was cheaper and faster but produced convulsions so violent that 42% of patients developed spinal fractures. In contrast, insulin was easier to control and less dangerous but took longer. But the debate would soon be moot, as both were replaced by treatments that were less risky *and* more effective.

Surgery. Surgery has been used for mental illness since ancient times, when trepanning, or cutting holes in the skull, was used to relieve pressure, release evil spirits, or perhaps both. But modern psychosurgery did not begin until 1936, when Portuguese physician Egas Moniz introduced the prefrontal leucotomy—the notorious lobotomy—in which an ice pick-like instrument was inserted into a patient's frontal lobes to sever their connections with other brain areas. The procedure seemed to work, and between 1935 and 1955 it was used in thousands of people and became a standard treatment for schizophrenia. Although Moniz won a Nobel Prize in 1949 for his technique, it eventually became clear that many patients were *not* helped and that its adverse effects included irreversible personality damage. After 1960, psychosurgery was modified to be less destructive, and today it is sometimes used for severe cases of mental illness.

Shock. In the late 1930s, Italian neurologist Ugo Cerletti was as impressed as anyone else when he heard that insulin and metrazol could improve the symptoms of schizophrenia. But given the risks, Cerletti thought he had a better idea. A specialist in epilepsy, he knew that electric shocks could cause convulsions, and so he joined with Italian psychiatrist Lucio Bini to develop a technique for delivering brief, controlled electric shocks. In 1938, after testing in animals, they tried their new electroconvulsive therapy, or ECT, on a delusional, incoherent man found wandering in the streets. The patient improved after just one treatment and recovered after a series of 11 treatments. Follow-up studies confirmed that ECT could improve schizophrenia, but physicians soon found it was even *more* effective for depression and bipolar disorder. Eventually, ECT replaced metrazol and insulin and became the preferred treatment around the world. Although ECT declined after the 1950s—due in part to concerns about its misuse—it was later refined and today is considered to be safe and effective for difficult-to-treat mental disorders.

And so by the 1940s, and for the first time in history, patients with serious mental disorders could be sickened, seized, severed, and shocked into feeling better. Not exactly confidence-inspiring, but enough of a milestone to encourage some researchers into believing they could find something better.

Milestone #2 Mastering mania: lithium takes on the "worst patient on the ward"

> "He is very restless, does not sleep, is irrational, and talks from one idea to another... He is so lacking in attention that questions fail to interrupt his flight of ideas... He is dirty and destructive, noisy both day and night... He has evidently been a great nuisance at home and in the neighborhood for the last few days..."
>
> —*From the case report of "WB," a patient with chronic mania*

This report—taken from the medical records of a patient who was about to change the course of medical history—illustrates how serious and troublesome mania can be, not only for patients, but anyone in their vicinity, inside or outside an institution. Although WB was in his 50s when these comments were written, the records of his mental illness date back from decades earlier, beginning just after he joined the Australian army in 1916 at the age of 21. Within a year, he would be hospitalized for "periods of permanent excitement" and discharged as medically unfit. Over the next few decades, he would be admitted to psychiatric hospitals multiple times with bouts of mania and depression, frustrating friends and family with behavior that ranged from "quiet and well behaved" to "mischievous, erratic, loquacious, and cunning." But perhaps his most memorable episode came in 1931, when he left a mental hospital dressed only in pajamas, entered a movie theater, and began singing to the audience.

By 1948, WB was in his 50s and had been a patient at Bundoora Repatriation Hospital in Melbourne, Australia, for five years. Diagnosed with chronic mania, the staff described him as "restless, dirty, destructive, mischievous, interfering, and long-regarded as the most troublesome patient on the ward." Small wonder that in March, 1948 physician John Cade chose WB as his first patient to try a new drug for treating mania—despite his initial impression that the drug would have the *opposite* effect.

Cade had begun looking for a treatment based on the theory that mania was a state of intoxication caused by some substance circulating in the blood. Figuring that the toxic substance might be found in the urine, he collected samples from patients with mania and injected it into animals. Cade's hunch was proved correct: The urine from patients with mania was more toxic than the urine from healthy people or those with other mental illnesses. He then began searching for the toxic substance in the urine. He soon narrowed his search to uric acid and isolated a particular form called "lithium urate." Perhaps he could treat mania by finding some way to block its effects. But to Cade's surprise, the compound had the opposite effect of what he expected. So he reversed his line of thinking: Perhaps lithium urate could *protect* against mania. After making a more pure form—lithium carbonate—he injected it into guinea pigs. When the animals responded with subdued behavior, Cade was sufficiently encouraged to try it out in people. After giving himself a dose to make sure it was safe, Cade administered the lithium to the most troublesome manic patient in the hospital…

And so on March 29, 1948, WB became the first patient in history to receive lithium for the treatment of mania. Almost immediately, WB began to "settle down," and a few weeks later, Cade was astonished to report, "There has been a remarkable improvement… He now appears to be quite normal. A diffident, pleasant, energetic little man." Two months after that, WB left the hospital for the first time in five years and was "soon working happily at his old job."

In addition to WB, Cade also gave lithium to nine other patients with mania, six with schizophrenia, and three with depression, but the effects in the patients with mania was particularly dramatic. He reported his findings in the *Medical Journal of Australia* the following year but left further studies of lithium to other Australian researchers, who conducted pivotal trials in the 1950s. Although lithium was not approved by the U.S. Food and Drug Administration until 1970, studies since then have showed that lithium use has significantly lowered mortality and suicidal behavior. In addition, one study found that between 1970 and 1991, the use of lithium in the U.S. saved more than $170 billion in direct and indirect costs. Lithium is far from perfect—it has many side effects, some potentially serious—but it still plays an important role today in the treatment of mania and several other mental illnesses.

With equal doses of curiosity and luck, Cade discovered the first effective drug for mental illness. His discovery was also a landmark because, in showing that lithium is more effective in mania than schizophrenia, he validated Emil Kraepelin's theory that the two disorders are distinct. It was the birth of a new understanding of mental illness and the start of a "Golden Age" of psychopharmacology that would continue for the next ten years with the discovery of three more milestone treatments.

Milestone #3 Silencing psychosis: chlorpromazine transforms patients *and* psychiatry

If you happened to be strolling down the streets of Paris in the early 1950s and bumped into a middle-aged man named Giovanni, you probably would have known exactly what to do: Cross quickly to the other side. Giovanni was a manual worker who often expressed himself in alarming ways, including giving impassioned political speeches in cafes, picking fights with strangers—and on at least one occasion—walking down the street with a flower pot on his head while proclaiming his love of liberty.

It's no surprise that in 1952, while a psychiatric patient in the Val-de-Grace Military Hospital in Paris, Giovanni was selected by doctors to try out a new drug. When the results were reported later in the year, the psychiatric community reacted with shock and disbelief. But within a few years, the drug—called chlorpromazine and better known as Thorazine in the U.S.—would be prescribed in tens of millions of patients around the world, literally transforming the treatment of mental illness.

Like many discoveries in medicine, the road to chlorpromazine was convoluted and unlikely, the result of a quest that initially had little to do with its final destination. Chlorpromazine was first synthesized in 1950 by scientists in France who were looking for a better antihistamine—not to cure allergic sniffles, but because they thought such drugs could help surgeons use lower doses of anesthesia and thus help patients better tolerate the trauma of surgery. In 1951, after initial studies suggested chlorpromazine might be a promising candidate, French anesthesiologist Henri-Marie Laborit administered it to his surgical patients at the Val-de-Grace hospital. Laborit was impressed *and* intrigued: Not only did the drug help patients feel better after their operations, it made them feel more relaxed and calm *before* the operation. Laborit speculated in an early article that there might be a "use for this compound in psychiatry."

By January, 1952, Laborit had convinced his colleagues in the hospital's neuropsychiatry department to try the drug in their psychotic patients. They agreed and found that giving chlorpromazine along with two other drugs quickly calmed down a patient with mania. But it was not until later that year that two other psychiatrists—Jean Delay and Pierre Deniker at the Sainte-Anne Hospital in Paris—tried giving chlorpromazine alone to Giovanni and 37 other patients. The results were dramatic: Within a day, Giovanni's behavior changed from erratic and uncontrollable to calm; after nine days he was joking with the medical staff and able to maintain a normal conversation; after three weeks he appeared normal enough to be discharged. The other psychotic patients showed a similar benefit.

Despite their initial shock, the psychiatric community was quick to embrace the new treatment. By the end of 1952, chlorpromazine was commercially available in France, and the U.S. followed with Thorazine in 1954. By 1955, studies around the world were confirming the therapeutic effects of chlorpromazine. Almost universally, psychiatrists were amazed at its effects in patients with schizophrenia. Within days, previously unmanageable patients—aggressive, destructive, and confused—were able to sit calmly with a clear mind, oriented to their surroundings, and talk rationally about their previous hallucinations and delirium. Clinicians reported that the atmosphere in mental hospitals changed almost literally overnight, as patients were not only freed from straitjackets, but from the institutions themselves.

By 1965, more than 50 million patients worldwide had received chlorpromazine, and the "deinstitutionalization" movement, for good or bad, was well underway. The impact was obvious in terms of shorter hospital stays and fewer admissions: A psychiatric hospital in Basel, Switzerland, reported that from 1950 to 1960 the average stay had decreased from 150 days to 95 days. In the United States, the number of patients admitted to psychiatric hospitals had increased in the first half of the twentieth century from 150,000 to 500,000; by 1975, the number had dropped to 200,000.

Although chlorpromazine was the most prescribed antipsychotic agent throughout the 1960s and 1970s, by 1990 more than 40 other antipsychotic drugs had been introduced worldwide. The push for new and better antipsychotics is understandable given the concerns about side effects. One study in the early 1960s found that nearly 40% of patients who took chlorpromazine or other antipsychotics experienced "extrapyramidal" side effects, a collection of serious symptoms that can include tremor, slurred speech, and involuntary muscle contractions. For this reason, researchers began to develop "second-generation" antipsychotics in the 1960s, which eventually led to the introduction of clozapine (Clozaril) in the U.S. in 1990. While clozapine and other second-generation agents pose their own risks, they're less likely to cause extrapyramidal symptoms. Second-generation psychotics are also better at treating "negative" symptoms of schizophrenia (that is, social withdrawal, apathy, and "flattened" mood), though none are superior to chlorpromazine for treating "positive" symptoms, such as hallucinations, delusions, disorganized speech.

Despite the dozens of antipsychotics available today, it's now clear that these drugs don't work in all patients, nor do they always address all of the symptoms of schizophrenia. Nevertheless, coming just a few years after the discovery of lithium, chlorpromazine was a major milestone. As the first effective drug for psychosis, it transformed the lives of millions of patients and helped reduce the stigma associated with mental illness. And so by the mid-1950s, drugs were now available for two major types of mental illness, bipolar disorder and schizophrenia. As for depression and anxiety, their moment in the sun was already on the horizon.

Milestone #4 Recovering the ability to laugh: the discovery of antidepressants

Most of us think we know something about depression—that painful sadness or "blues" that periodically haunts our lives and sets us back for a few hours or days.

Most of us would be wrong.

True clinical depression—one of the four most serious mental disorders and projected to be the second-most serious burden of disease in 2030—is not so much a setback as a tidal wave that overwhelms a person's ability to *live*. When major depression gets a grip, it does not let go, draining energy, stealing interest in almost all activities, blowing apart sleep and appetite, blanketing thoughts in a fog, hounding a person with feelings of worthlessness and guilt, filling them with obsessions of suicide and death. As if all that weren't enough, up until the 1950s, people with depression faced yet one *more* burden: the widespread view that the suffering was their own fault, a personality flaw that might be relieved with psychoanalysis, but certainly not drugs. But in the 1950s, the discovery of two drugs turned that view on its head. They were called "antidepressants"—which made a lot more sense than naming them after the conditions for which they were originally developed: tuberculosis and psychosis.

The story of antidepressants began with a failure. In the early 1950s, despite Selman Waksman's recent milestone discovery of streptomycin—the first successful antibiotic for tuberculosis (see Chapter 7)—some patients were not helped by the new drug. In 1952, while searching for other anti-tuberculosis drugs, researchers found a promising new candidate called iproniazid. In fact, iproniazid was *more* than promising, as seen that year in dramatic reports from Sea View Hospital on Staten Island, New York. Doctors had given iproniazid to a group of patients who were dying from tuberculosis, despite being treated with streptomycin. To everyone's shock, the new drug did more than improve their lung infections. As documented in magazine and news articles that caught the attention of the world, previously terminal patients were "reenergized" by iproniazid, with one famous photograph showing some "dancing in the halls." But although many psychiatrists were impressed by iproniazid and considered giving it to their depressed patients, interest soon faded due to concerns about its side effects.

Although reports that iproniazid and another anti-tuberculosis drug (isoniazid) could relieve depression were sufficiently intriguing for American psychiatrist Max Lurie to coin the term "antidepressant" in 1952, it was not until a few years later that other researchers began taking a harder look at iproniazid. The landmark moment finally arrived in April 1957, when psychiatrist Nathan Kline reported at a meeting of the American Psychiatric Association that he'd given iproniazid to a small group of his depressed patients. The results were impressive: 70% had a substantial improvement in mood and other symptoms. When additional encouraging studies were presented at a symposium later that year, enthusiasm exploded. By 1958—despite still being marketed only for tuberculosis—iproniazid had been given to more than 400,000 patients with depression.

While researchers soon developed other drugs similar to iproniazid (broadly known as MAOIs), all shared the same safety and side effect issues seen with iproniazid. But before long, thanks to the influence of chlorpromazine, they found an entirely new kind of antidepressant.

In 1954, Swiss psychiatrist Roland Kuhn, facing a tight budget at his hospital, asked Geigy Pharmaceuticals in Basle if they had any drugs he could try in his schizophrenic patients. Geigy sent Kuhn an experimental compound that was similar to chlorpromazine. But the drug, called G-22355, not only failed to help his psychotic patients, some actually became *more* agitated and disorganized. The study was dropped, but upon further review, a curious finding turned up: three patients with "depressive psychosis" had actually *improved* after the treatment. Suspecting that G-22355 might have an antidepressant effect, Kuhn gave it to 37 patients with depression. Within three weeks, most of their symptoms had cleared up.

As Kuhn later recounted, the effects of this new drug were dramatic: "The patients got up in the morning voluntarily, they spoke in louder voices, with great fluency, and their facial expressions became more lively... They once more sought to make contact with other people, they began to become happier and to recover their ability to laugh."

The drug was named imipramine (marketed in the U.S. as Tofranil), and it became the first of a new class of antidepressants known as tricyclic antidepressants, or TCAs. After the introduction of imipramine, many other TCA drugs were developed in the 1960s. Lauded for their relative safety, TCAs soon became widely used, while MAOIs fell out of favor. But despite their popularity, TCAs had some safety concerns, including being potentially fatal if taken in overdose, and a long list of side effects.

The final stage in the discovery of antidepressants began in the 1960s with yet another new class of drugs called SSRIs. With their more targeted effects on a specific group of neurons—those that release the neurotransmitter serotonin—SSRIs promised to be safer and have fewer side effects than MAOIs or TCAs. However, it wasn't until 1974 that one particular SSRI was first mentioned in a scientific publication. Developed by Ray Fuller, David Wong, and others at Eli Lilly, it was called fluoxetine; and in 1987, it became the first approved SSRI in the U.S., with its now-famous name, Prozac. The introduction of Prozac—which was as effective as TCAs but safer and comparatively free of side effects—was the tipping point in the milestone discovery of antidepressants. By 1990, it was the most prescribed psychiatric drug in North America, and by 1994, it was outselling every drug in the world except Zantac. Since then, many other SRRIs and related drugs have been introduced and found to be effective in depression.

The discovery of iproniazid and imipramine in the 1950s was a major milestone for several reasons. In addition to being the first effective drugs for depression, they opened a new biological understanding of mood disorders, prompting researchers to look at the microscopic level of where these drugs work, leading to new theories of how deficiencies or excesses of neurotransmitters in the brain may contribute to depression. At the same time, the new drugs literally transformed our understanding of what depression is and how it could be treated. Up until the late 1950s, most psychiatrists believed in the Freudian doctrine that depression was not so much a "biological" disorder as a psychological manifestation of internal personality conflicts and subconscious mental blocks that could only be resolved with psychotherapy. Many outwardly resisted the idea of drug treatments, believing they could only mask the underlying problem. The discovery of antidepressants forced psychiatrists to see depression as biological disorder, treatable with drugs that modified some underlying chemical imbalance.

Today, despite many advances in neurobiology, our understanding of depression and antidepressant drugs remains incomplete: We still don't know exactly how antidepressants work or why they don't work *at all* in up to 25% of patients. What's more, studies have shown that in some patients psychotherapy can be as effective as drugs, suggesting that the boundaries between biology and psychology are not clear cut. Thus, most clinicians believe that the best approach to treating depression is a combination of antidepressant drugs *and* psychotherapy.

Apart from their impact on patients and psychiatry, the discovery of antidepressants in the 1950s had a profound impact on society. The fact that these drugs can dramatically improve the symptoms of depression—yet have little effect on people with normal mood states—helped society realize that clinical depression arises from biological vulnerabilities and not a moral failing or weakness in the patient. This has helped destigmatize depression, placing it among other "medical" diseases—and apart from the "blues" that we all experience from time to time.

Milestone #5 More than "mother's little helper": a safer and better way to treat anxiety

Anxiety is surely the least serious of the four major mental disorders: It goes away when the "crisis" is over, has simple symptoms compared to bipolar disorder or schizophrenia, and we've *always* had plenty of treatments, from barbiturates and other chemical concoctions, to the timeless remedies of alcohol and opium. In short, anxiety disorders are not really as serious as the other major forms of mental illness, are they?

They are. First, anxiety disorders are by far the most common mental disorders, affecting nearly 20% of American adults (versus about 2.5% for bipolar disorder, 1% for schizophrenia, and 7% for depression). Second, they can be as disabling as any mental disorder, with a complex of symptoms that can be mental (irrational and paralyzing fears), behavioral (avoidance and quirky compulsions), *and* physical (pounding heart, trembling, dizziness, dry mouth, and nausea). Third, anxiety disorders are as mysterious as any other mental disorder, from their persistence and resistance to treatment, to the fact that they can crop up in almost any other mental disorders, including—paradoxically—depression. And finally, prior to the 1950s, virtually all treatments for anxiety posed a risk of three disturbing side effects: dependence, addiction, and/or death.

The discovery of drugs for anxiety began in the late-1940s when microbiologist Frank Berger was looking for a drug *not* to treat anxiety, but as a way to preserve penicillin. Berger was working in England at the time and had been impressed by the recent purification of penicillin by Florey and Chain (see Chapter 7). But a funny thing happened when he began looking at one potential new preservative called mephenesin. While testing its toxicity in laboratory animals, he noticed that it had a *tranquilizing* effect. Berger was intrigued, but the drug's effects wore off too quickly. So, after moving to the U.S., he and his associates began tweaking the drug to make it last longer. In 1950, after synthesizing hundreds of compounds, they created a new drug that not only lasted longer, but was eight *times* more potent. They called it meprobamate, and Berger was optimistic about its therapeutic potential: It not only relieved anxiety, but relaxed muscles, induced mild euphoria, and provided "inner peace."

Unfortunately, Berger's bosses at Wallace Laboratories were less impressed. There was no existing market for anti-anxiety drugs at the time, and a poll of physicians found that they were not interested in prescribing such drugs. But things changed quickly after Berger put on his marketing cap and made a simple film. The film showed rhesus monkeys under three conditions: 1) their naturally hostile state; 2) knocked out by barbiturates; and 3) calm and awake while on meprobamate. The message was clear, and Berger soon garnered the support he needed. In 1955 meprobamate was introduced as Miltown (named after a small village in New Jersey where the production plant was located), and once word of its effects began to spread, it quickly changed the world.

And the world was more than ready. Although barbiturates were commonly used in the early 1950s, the risk of dependence and fatality if taken in overdose were well-known. At the same time, recent social changes had prepared society to accept the idea of a drug for anxiety, from a growing trust in the pharmaceutical industry thanks to the recent discovery of penicillin and chlorpromazine, to the widespread angst about nuclear war, to the new work pressures brought on by the economic expansion following World War II. Although some claimed that meprobamate was promoted to exploit stressed housewives—leading to the cynical nickname "mother's little helper"—Miltown was widely used by men and women around the world, including business people, doctors, and celebrities. By 1957, more than 35 million prescriptions had been sold in the U.S., and it became one of the top 10 best-selling drugs for years.

But while doctors initially insisted meprobamate was completely safe, reports soon began to emerge that it could be habit-forming and—though not as dangerous as barbiturates—potentially lethal in overdose. Soon, pharmaceutical companies were looking for safer drugs, and it didn't take long. In 1957, Roche chemist Leo Sternbach was cleaning up his laboratory when an assistant came across an old compound that had never been fully tested. Sternbach figured it might be worth a second look, and once again random luck paid off. The drug not only had fewer side effects than meprobamate, but was far more potent. It was called chlordiazepoxide, and it became the first of a new class of anti-anxiety drugs known as benzodiazepines. It was soon marketed as Librium and would be followed in 1963 by diazepam (Valium) and many others, including alprazolam (Xanax). By the 1970s, benzodiazepines had mostly replaced meprobamate and began to play an increasingly important role in treating anxiety disorders.

Today, in addition to benzodiazepines, many other drugs have been found to be useful in treating anxiety disorders, including various antidepressants (MAOIs, TCAs, and SSRIs). While today's benzodiazepines still have limitations—including the risk of dependence if taken long-term—they are considered far safer than the drugs used up until the 1950s, before Frank Berger began his quest for a way to preserve penicillin.

While some have criticized the widespread use of anti-anxiety agents, it would be foolish to dismiss the remarkable benefits these drugs have provided for millions of people who lives would otherwise be crippled by serious anxiety disorders. What's more, similar to the earlier milestone drugs for mental illness, their discovery opened new windows into the study of normal brain functions and the cellular and molecular changes that underlie various states of anxiety. This, in turn, has advanced our understanding of the biological underpinnings of the mind and thus helped destigmatize mental illness.

How four new treatments added up to one major breakthrough

The 1950s breakthrough discovery of drugs for madness, sadness, and fear was both a "Golden Age" of psychopharmacology and a transformative awakening of the human race. First and foremost, the new drugs helped rescue countless patients from immeasurable suffering and loss. To the shock of almost everyone, they helped patients regain their ability to think and act rationally, to laugh and talk again, to be freed from irrational and crippling fears. Patients could resume relationships and productive lives or simply stay alive by rejecting suicide. Today, NAMI estimates that a combination of drugs and psychosocial therapy can significantly improve the symptoms and quality of life in *70% to 90%* of patients with serious mental illness. With millions of lives saved by various drugs for mental illness, it's no wonder that many historians rank their discovery as equal to that of antibiotics, vaccines, and the other top ten medical breakthroughs.

But almost as important as their impact on patients was the way in which these drugs transformed the prejudices and misconceptions long held by patients, families, doctors, and society. Prior to the 1950s, mental illness was often viewed as arising from internal psychological conflicts, somehow separate from the mushy biology of the brain and suspiciously linked to an individual's personal failings. The discovery that specific drugs relieved specific symptoms implicated biochemical imbalances as the guilty party, shifting the blame from "slacker" patients to their "broken" brains.

Yet for all of the benefits of drugs for mental disorders, today most of the same mysteries remain, including what causes mental illness, why the same symptoms can appear in different conditions, why some drugs work for multiple disorders, and why they sometimes don't work at all. What's more, while pharmacologists continue their unending quest for better drugs and new explanations, one underlying truth seems unlikely to ever change: Drugs alone will never be sufficient...

The failures of success: a key lesson in the treatment of mental disorders

> "There has been a remarkable improvement... He now appears to be quite normal. A diffident, pleasant, energetic little man."
>
> —*John Cade, after the first use of lithium in WB, a patient with mania*

You'll recall this earlier comment from the remarkable story of WB, who after receiving lithium in 1948 for his mania became the first person to be successfully treated with a drug for mental illness. Unfortunately, that story was a lie. Unless, that is, the remaining details of WB's life are revealed. WB *did* have a dramatic recovery after his treatment with lithium and he continued to do well for the next six months. But trouble began a short time later when, according to case notes, WB "stopped his lithium." A few days later, his son-in-law wrote that WB had returned to his old ways, becoming excitable and argumentative after a trivial disagreement. Over the next two years, WB repeatedly stopped and restarted his lithium treatment, causing his behavior to zig-zag from "irritable, sleepless, and restless," to "normal again," to "noisy, dirty, mischievous, and destructive as ever." Finally, two years after his milestone treatment, WB collapsed in a fit of seizures and coma. He died a few miserable days later, the cause of death listed as a combination of lithium toxicity, chronic mania, exhaustion, and malnutrition.

The full story of WB highlights why the breakthrough discovery of drugs for mental illness was both invaluable and inadequate. WB's decline was not simply due to his failure to take his lithium, but problems with side effects, proper dosing, and even the self-satisfied symptoms of mania itself. All of these problems, in one form or another, are common in many mental disorders and a major reason why treatment can fail. As society learned during the deinstitutionalization movement of the 1960s and 1970s, drugs can produce startling improvements in mental functioning but are sometimes woeful failures in helping patients negotiate such mundane challenges as maintaining their treatment, finding a job, or locating a place to live.

Sadly, the same story continues today, sometimes to the detriment of both patients and innocent bystanders. When David Tarloff's deterioration from schizophrenia led to the murder of psychologist Kathryn Faughey in 2008, it followed years of being on and off various anti-manic and antipsychotic medications—including lithium, Haldol, Zyprexa, and Seroquel—and being admitted and released from more than a dozen mental institutions. As Tarloff's brother lamented to reporters after David's arrest, "My father and I and our mother tried our best to keep him in the facility he was hospitalized in over the many years of his illness, but they kept on releasing him. We kept asking them to keep him there. They didn't..."

In 2008, the World Health Organization issued a report on mental health care, noting that of the "hundreds of millions of people" in the world with mental disorders, "only a small minority receive even the most basic treatment." The report advised that this treatment gap could be best addressed by integrating mental health services into primary care, with a network of supporting services that include not only the drugs described here, but collaboration with various informal community care services. It is these community services—traditional healers, teachers, police, family, and others—that "can help prevent relapses among people who have been discharged from hospitals."

The treatment of mental illness has come a long way since Hippocrates prescribed laxatives and emetics, since asylum directors attempted to "master" madness with chains and beatings, since doctors sickened their patients with malaria and seizures. The breakthrough discoveries of the first drugs for mental illness changed the world, but also laid bare some timeless truths. As the stories of David Tarloff and WB remind us, the safety net that can catch and support patients must be woven from many strands—medications, primary care and mental health practitioners, community, and family—the absence or failure of any one of which can lead to a long, hard fall.

A Return to Tradition: THE REDISCOVERY OF ALTERNATIVE MEDICINE 10

Case 1: A bad day for Western medicine

The year was 1937, and pneumonia season was in full swing. The sick ward at Boston City Hospital—a large, open room with 30 beds neatly arranged around its perimeter—was filling rapidly with patients suffering from its telltale symptoms: chills, fever, bloody cough, and pain in one side of the chest. Yet one patient, a young black musician, stubbornly refused to cooperate. Admitted a few days earlier with chills and fever, he had no cough. And now, as a young intern fresh out of Harvard Medical School made his way through the ward and stopped at the man's bedside, the patient was unable to produce the sputum sample needed to diagnose pneumonia. The intern, Lewis Thomas, was in just the first month of his service and didn't think much of it at the time. Drawing the requisite blood sample, he moved on. It was not until later that morning in the upstairs laboratory that Thomas peered through a microscope at the blood sample and made his astonishing discovery.

Thomas immediately notified the hospital's hematologists, who rushed up to the laboratory to take a look, then charged back down to the ward to collect their own samples. Soon, word spread like wildfire through the hospital, as staff doctors, visiting physicians, and students all raced to the ward and then hurried upstairs to view the evidence with their own eyes. The patient did not have pneumonia, but *malaria*, a deadly parasitic disease transmitted to humans by infected mosquitoes. Once inside a human, the parasites can invade red blood cells, reproduce, and cause them to rupture. This is what Thomas saw on the microscope slide, and who wouldn't be transfixed by the sight? Red blood cells literally bursting open, with tiny parasites swarming out and seeking other cells to infect. The hospital-wide sensation was only heightened by an additional mystery: Malaria normally occurs in the tropics and subtropics; how could a person in the northern climes of Boston—with no recent travel outside the country—be infected?

Although the mystery was soon solved—the patient was a heroin addict and probably contracted the disease from a used syringe contaminated by someone from out of town—the fascination continued. It continued as doctors visited the ward all afternoon, continued as they drew more blood to look at more slides, continued into the evening as the patient grew weaker, comatose—and suddenly died. In fact, the doctors were so fascinated with the disease, they forgot to treat the *patient*. As

Thomas sadly recalled years later in his book, *The Youngest Science*, if instead of "animated attention" the patient had been immediately treated with quinine—a malarial drug whose curative powers had been known since the *seventeenth century*—then "he would perhaps have lived. The opportunity to cure an illness, even save a life, came infrequently enough on the City Hospital wards. This one had come and gone.... It was a bad day for Harvard."

Case 2: A bad day for Eastern medicine

The year is 2008, and the terror begins just minutes after Yonten sits down to meditate. For Yonten, a 45-year-old Tibetan monk, what haunts him most is not the memory of angry shouting, nor the Chinese authorities torturing his fellow monks with electric shocks, nor even the prison beating he underwent just before his recent escape from Tibet. Rather, it is the image of his monastery going up in flames that burns most vividly in his mind. And despite the decades he has spent mastering Buddhist meditation, none of Yonten's skills—deep breathing, energy channeling, focused mindfulness—can remove the image of the burning temple that shatters his attempts to meditate. In fact, the harder he tries, the greater his frustration, and instead of inner peace, he is rewarded with feelings of sadness, guilt, and hopelessness.

Although now living safely in the United States, Yonten and many other refugee Tibetan monks continue to be haunted by memories of torture and abuse that disrupt their ability to meditate and practice their religion. The good news is that Tibetan traditional healers have diagnosed their condition as *srog-rLung*, or "life-wind" imbalance. The idea that balance is essential to health—and that imbalance leads to disease—is hardly unique to Tibetan medicine. Many ancient healing traditions dating back thousands of years teach that the human body is inseparable from the outer world and interconnected by invisible forces. According to these traditions, the secret to good health is maintaining a *balance* among these inner and outer forces. The bad news for Yonten and his fellow monks is that among the many traditional treatments used to help restore balance, the one they're using is not working. In fact, it's failing miserably.

Meditation is an Eastern healing tradition that dates back thousands of years, crosses many cultural boundaries, and today is one of the top three forms of alternative medicine in the United States. For Buddhist

Tibetan monks, meditation is the *ultimate* treatment, a method of attaining Enlightenment, which they view as the cure for all suffering. Yet for Yonten and other refugee monks traumatized by their experiences in Tibet, the meditation skills they have spent a lifetime mastering are not merely failing, but *causing* numerous symptoms, from guilt and depression to elevated blood pressure and heart palpitations. The problem is the treatment itself: The form of meditation they're practicing is so "single-minded," it violates their *own* principles of balance.

Lessons from the past: East meets West (again)

Two stories of treatment failure from opposite ends of the cultural spectrum: In a world of Western scientific medicine, a young musician dies of malaria when hospital physicians become so fascinated by a disease, they forget to treat the *patient*. In a world of traditional Eastern medicine, Tibetan monks are so haunted by memories of torture and abuse that the meditation skills they have mastered to prevent suffering are now *causing* it.

These two stories symbolize how medicine, regardless of its cultural origin, can become a victim of its own methods, even its own success. But more than that, they are a starting point for a much larger story of how two medical traditions, born thousands of years ago from common roots, became divided for centuries, fought an ugly battle over differences in philosophy, and then finally rejoined forces at the brink of the twenty-first century to become one of the top ten breakthroughs in medicine.

* * *

The accidental death of a young black musician from malaria in 1937 may seem an isolated event, but it represented an ominous trend that was beginning to emerge in twentieth-century medicine. Thanks to a growing list of breakthroughs—vaccines, germ theory, anesthesia, X-rays, and many more—scientific medicine was establishing itself as *the* dominant medical system of the Western world. But with those breakthroughs came a hidden malaise: Enamored by new technologies and an explosion of information, medicine was becoming distracted, forgetting that its primary focus was not disease, but the *patient*; that while it was not always possible to cure, it was always essential to *care*.

What caused this shift in priorities? The rise of modern medicine in the twentieth century may seem inevitable, but it did not have to be this way. Barely 100 years earlier, scientific medicine—also sometimes called conventional medicine or biomedicine—was just another *alternative* medicine, one of many approaches to healthcare up for grabs at the time. In fact, up until the late 1800s, "scientific" medicine was often a barbaric and risky business, with its crude surgery, blood-letting, and use of toxic drugs like mercury as laxatives and emetics. At the time, many other healing systems were competing with scientific medicine for legitimacy if not dominance, including hydrotherapy (use of hot and cold water to prevent and treat disease), Thomsonianism (mix of Native American herbal therapy and medical botanical lore), and magnetic therapy (use of healing touch to transfer "magnetic" or "vital" energy into a patient). For decades, each viewed the other with mistrust and disdain, exchanging cross-charges of quackery and malpractice. It was not until Western scientific medicine began to exert its dominance in the late 1800s that other healthcare models gradually fell from popularity and became, by default, "alternative."

It's no mystery why scientific medicine won the initial battle. With its emphasis on experimentation, observation, and reason—the so-called "scientific method" that blossomed in the eighteenth and nineteenth centuries—scientific medicine had found a potent way to explore and explain the world. But perhaps most significant, it led scientists down a rabbit hole of *reductionism*, of increasingly dividing the body into smaller and smaller parts. With powerful new tools such as microscopes, X-rays, and various laboratory techniques, scientists began delving deeper and deeper into the mysteries of tissues, organs, cells, and beyond, revealing stunning secrets of physiology and disease, each discovery seeming to lead to new treatments.

But as the pace of discoveries accelerated into the twentieth century, the balance of medicine began to shift: Technology and specialization changed the way doctors looked at patients, turning them into a collection of parts and diseases instead of the unique "whole" individuals who had walked into their office seeking health*care*. By the 1980s, the use of managed care to control the burgeoning costs of technology left even *less* time for physicians and their patients, and further degraded patients by turning them into disease categories. By the final decades of the twentieth century, scientific medicine—despite remarkable successes ranging from organ transplants to heart surgery and cancer treatment—had lost its balance, unleashing a backlash of frustration so strong that more and more patients were demanding an *alternative*.

As it turned out, the alternative had never gone away.

Despite the dominance of scientific medicine in the twentieth century, many alternative therapies born in the previous century—including chiropractic, osteopathic, and homeopathic medicine—continued to survive and evolve. As many patients turned to these options in the 1970s and 1980s, others looked to ancient alternatives, including traditional Chinese medicine and Indian Ayurvedic medicine, which not only offered entire medical systems, but specific treatments such as meditation, massage, and acupuncture. In the end, the bottom line was simple. Alternative medicine offered something Western medicine had too often abandoned: the view that every patient was an individual; that natural treatments were sometimes better than dramatic surgery and dangerous drugs; and that the essence of medicine *begins* with a caring relationship between healer and patient.

The reaction by modern medicine to this trend was predictable: the same denial and derision with which it had regarded alternative medicine for the previous 150 years. But in the late 1990s, the milestone moment finally arrived in a form that scientific medicine could not afford to ignore: Two of its most prestigious medical journals—the *New England Journal of Medicine* and the *Journal of the American Medical Association*—reported that not only had alternative medicine use been increasing "dramatically," but that by 1998, Americans were actually seeking alternative care practitioners *more often* than their own primary care physicians.

The wake-up call had been sounded, and one of the top ten breakthroughs in medicine had arrived: the *re*discovery of alternative medicine. But the full story of this breakthrough reached much further back in time than the past few decades or even centuries. In fact, its roots can be traced back through almost every stage of medical history—from the rise of traditional medicine at the dawn of civilization, to a revolutionary change during the Renaissance; from the birth of "alternative" medicine in the 1800s, to the twentieth century battles that led to the rediscovery of alternative medicine—*and* something more.

Milestone #1

The birth of traditional medicine: when caring was curing

They emerged from the misty origins of civilization thousands of years ago and at first glance seemed to have almost nothing in common. Yet despite vast differences in geography, culture, and language, three of the first major systems of ancient medicine—traditional Chinese medicine, Indian Ayurvedic medicine, and Greek Hippocratic medicine—shared some remarkable similarities. It's not just that they all arose out of legend and magical/religious practices several thousand years ago and developed into their classical forms around 600 to 300 BC. Rather, it is that all three discovered some of the most important principles medicine would ever know—and one day, *forget*.

Traditional Chinese medicine

Born about 5,000 years ago in ancient China, Huang-Di must have been a very busy man during his reputed 100 years of life: In addition to founding the Chinese civilization, he is said to have: taught the Chinese how to build houses, boats, and carts; invented the bow and arrow, chopsticks, ceramics, writing, and money; and still somehow found time to father no less than 25 children. But Huang-Di, also known as the Yellow Emperor, is also famous for one *other* major milestone in Chinese history: discovering the principles of traditional Chinese medicine. Although his text, *Huang Di Nei Jing* (*Inner Canon of the Yellow Emperor*), was probably not compiled until thousands of years after his death (around 300 BC), today it remains a classic in traditional Chinese medicine, from its early descriptions of acupuncture to its ancient theories of physiology, pathology, diagnosis, and treatment.

But perhaps even more important, the *Inner Canon* introduced to traditional Chinese medicine (TCM) the philosophy of Taoism, with its two major teachings: First, that the human body is a microcosm of the universe and, thus, interconnected with nature and its forces; and second, that health and disease are determined by the *balance* of forces within the body and its connection to the outer world. The *Inner Canon* also describes many other key concepts in TCM, such as the theories of *yin-yang* (the world is shaped by two opposing yet complementary forces); *qi* (a vital energy or life force that circulates through the body in a system of pathways called meridians); the five elements (the relation of fire, earth, metal, water, and wood to specific organs in the body and their functions); and the "eight principles" used to analyze symptoms and categorize disease (cold/heat, interior/exterior, excess/deficiency, and yin/yang).

Yet despite the many forms of treatment provided by TCM—including herbs, acupuncture, massage, and movement therapies such as tai chi and qi gong—two underlying principles stand out:

1. Treatment is designed to help patients restore the balance of their qi, or vital energy.
2. Treatment is individualized based on the healer's careful evaluation of the patient, using such traditional methods as detailed observation, questioning, hearing/smelling, and touching/palpating.

Indian Ayurvedic medicine

Ayurvedic medicine can also be traced back to busy times around 5,000 years ago when, according to one legend, a group of sages gathered in the Himalayas to stop an ongoing epidemic of disease and death. In this lofty setting, the god Brahma taught the art of healing to Daksha, who taught it to Indra, who taught it to Bharadvaja, who taught it to Atreya, who taught it to six disciples, who—finally—compiled the knowledge into the Ayurveda. No word on what happened in the meantime with the epidemic. Legends aside, modern scholars generally trace Ayurvedic medicine back to at least 1,000 BC, when an early form known as Atharavaveda was dominated by magical/religious practices. However, similar to TCM, around 500 to 300 BC, a new classical form arose that combined past knowledge with new ideas. It was called Ayurveda, or "science of life," from the Sanskrit words *ayur* (life) and *veda* (science).

Despite some obvious differences, Ayurvedic medicine is remarkably similar to traditional Chinese medicine in its basic philosophy, including the view that all living and nonliving things in the universe are interconnected and that disease arises when a person is out of balance with the universe. At the same time, Ayurvedic medicine has its own unique terminology and ideas, including the idea that each person has a unique *prakriti*, or constitution, which in turn, is influenced by three *doshas* (life energies). While the system is complex, one underlying message, similar to TCM, is that disease can arise if there is an imbalance in a particular *dosha*. Ayurvedic medicine also shares the same patient-focused approach seen with other traditional medicines, including a detailed and sophisticated system for examining patients. Once the nature of the illness is determined, treatment is based on a variety of individualized therapies, such as herbs, massage, breathing exercises, meditation, and dietary changes. And while some treatment goals are unique to Ayurvedic medicine, the ultimate goal is unmistakably familiar: to restore health by improving *balance* in the patient's body, mind, and spirit.

Greek Hippocratic medicine

When we last visited with Hippocrates for any length of time, he had just achieved one of the top ten breakthroughs in the history of medicine: the discovery of medicine itself (Chapter 1). Indeed, even as classical traditional medicine was evolving in China and India, the milestone achievements by Hippocrates and his followers were defining the profession of medicine itself. Yet Hippocratic medicine was also a *traditional* medicine, with many similarities to early Chinese and Indian medicine. For example, the roots of Hippocratic medicine can also be traced to 1,000 BC or earlier, when medicine was practiced in the Asklepieion healing temple on the island of Kos (see Chapter 1).

But by the time ancient Greek medicine developed into its classical form in the fifth century, Hippocrates was teaching many concepts similar to the classical Chinese and Ayurvedic medicine emerging at the time, including the view that health was influenced by interactions among the body, mind, and environment. Of course, Hippocratic medicine had its *own* unique system, including the belief that the body produced four circulating fluids, or humors—blood, phlegm, yellow bile, and black bile. Nevertheless, similar to the other traditions, Hippocrates taught that disease arose from some imbalance—either among the patients' humors or in their relation to the outer world—and that the goal of treatment was to restore a healthy balance. Hippocratic medicine also shared a similar approach to treatment seen in other ancient traditions, including the use of remedies such as dietary restrictions, exercise, and herbs. In addition, Hippocrates strongly emphasized the physician-patient relationship, teaching that the best way to diagnose and predict the course of disease was through lengthy interviews, careful observation, and detailed examinations. This emphasis was documented in excruciating detail in the text *Epidemics 1*, which teaches that physicians should not only study "the common nature of all things" but also the patient's "customs, way of life, age, talk, manner, silence, thoughts, sleeping, dreams, plucking/scratching/tearing, stools, urine, sputa, vomit, sweat, chill, coughs, sneezes, hiccups, flatulence, hemorrhoids, and bleeding."

* * *

Despite some obvious differences, the earliest medical traditions uncovered the same secrets of health and disease, from the interconnectedness of the body, mind, spirit, and universe to the importance of balance and natural treatments. What's more, all emphasized the patient-healer relationship and patient *care*. With such principles, it's no surprise that traditional Chinese medicine and Ayurvedic medicine would thrive for the next 2,500 years. But though Hippocratic medicine would remain influential for more than a millennium, beginning around the sixteenth century, a revolutionary change would send it down a different path and an entirely new way of looking at the world.

Milestone #2 Enlightenment: 1,200 years of tradition overturned and a new course for medicine

It is perhaps the greatest irony in the history of medicine: A Greek physician whose brilliance was second only to Hippocrates and whose discoveries and writings were influential for more than 1,000 years is more often remembered today for his biggest *mistakes*. Yet when two individuals discovered during the Renaissance that Galen had made a number of critical errors, they not only overturned a long tradition of misinformation—they *also* gave birth to a new world of modern scientific medicine.

Born in Pergamum (now modern-day Turkey) in 129 AD, Galen's skills as a physician were so admired that he was made physician to the son of the famous Roman emperor Marcus Aurelius. However, it was his numerous discoveries in anatomy and physiology, along with tracts on medicine and ethics, which earned him fame and influence for more than a millennium. With a passion for truth that bordered on arrogance—he once wrote, "My father taught me to despise the opinion and esteem of others and to seek only the truth"—Galen investigated every branch of medicine possible at the time and became famous for his skills as a physician, his animal dissections, and lectures. Among his many great discoveries was the finding that arteries carry blood rather than air and that muscles are controlled by nerves coming from the brain. Unfortunately, Galen also held many false beliefs, particularly his view that the liver—and not the heart—was the central organ of the circulatory system. And so, just as Galen's brilliant insights were passed on for 1,200 years, so were many of his misconceptions.

It was not until the Renaissance that people began to question ancient writings long-assumed to be true. During this time, many great thinkers were beginning to change how the world was viewed, such as Nicolaus Copernicus' theory in the 1500s that the earth revolved around the sun, rather than vice versa. However, the greatest transformation in medicine began with the work of two physicians—Andreas Vesalius and William Harvey—whose groundbreaking study of the human body overturned tradition and set medicine on a revolutionary new course.

Andreas Vesalius was a Belgian physician and anatomist who was born in 1514 and who, as a child, not only enjoyed dissecting small animals, but also the bodies of executed criminals left out in the open on some land near his family home. It was just the kind of hands-on experience he needed to realize, by the time he had finished his medical training and had been appointed professor of surgery and anatomy in Padua, Italy, that what he had been taught as a student did not match what he'd personally seen in his own dissections. And so after his medical training, Vesalius continued dissecting cadavers and was soon admired not only for his careful and detailed dissections, but his lectures and demonstrations. Although Vesalius initially performed his dissections in an attempt to account for the differences with the writings of Galen, he eventually began to lose faith: He uncovered more than 200 errors in Galen's work, including Galen's belief that the human jawbone has two sections (it only has one) and that there is a coil of blood vessels at the base of the human brain (there isn't). While many of the mistakes were understandable—Galen had dissected animals, while Vesalius worked on human cadavers—it did not stop Vesalius from setting the record straight.

And so in 1543, at the age of 29, Vesalius published a seven-volume work, *De Humani Corporis Fabrica*, capturing his years of masterful dissection work. Comprised of more than 300 finely detailed illustrations of human anatomy, it was the first book of its kind and immediately recognized as a masterpiece. While some resisted the idea that Vesalius' work contradicted the long-revered texts of Galen, the unprecedented detail and proof of Vesalius' work spoke for itself. In exposing the errors of Galen, the *Fabrica* set a new standard that subsequent generations would not forget: Detailed observation and recorded facts *must* take precedence over unexamined assumptions.

While Vesalius had exposed Galen's mistakes in anatomy, just a few decades later English physician William Harvey pursued his own trail of truth to uncover equally shocking mistakes in physiology. Until that time, scientists had not questioned Galen's explanation of how blood flowed through the body. For example, Galen had taught that blood, rather than circulating continuously through the body by the pumping heart, was continuously created in the liver, pushed through the body by an "ebb and flow" of the heart, and sent to the tissues where it was "consumed." Galen also thought that once in the heart, blood passed through pores in the wall between its lower chambers (ventricles). But Harvey, who was born in 1578 and raised in a time when his idols, including Vesalius, promoted experimentation, decided to take a closer look.

And in 1616, after numerous experiments on a variety of animals, Harvey announced his stunning discovery to the world: "The blood moves as in a circle. The arteries are the vessels carrying blood from the heart to the body and the veins returning blood from the body to the heart." It was a completely different concept than that described by Galen and, although Harvey faced some criticism and doubters, he eventually published his findings in 1628 in the small book, *De Motu Cordis*. In addition to accurate descriptions of how the heart receives and pumps blood to the body, he correctly described the different functions of veins and arteries, and in one famous contradiction to Galen's work, concluded that blood does not flow through the wall in the heart "because there are no openings."

* * *

It may sound simple today, but some call Harvey's explanation of circulation the greatest discovery in physiology and medicine. What's more, like Vesalius' milestone revelations in human anatomy, Harvey's discovery transcended the mere embarrassment of a biological boo-boo. Coming after more than *1,200 years* of unquestioned authority, Vesalius and

Harvey dared look at the human body in a way no one had before. Overturning tradition, they threw open doors to a new way of seeing the world. It was the start of a slow transformation that medicine—*scientific* medicine—would undergo over the next five centuries.

Milestone #3 The birth of scientific medicine: when curing began to overshadow caring

While the revolutionary work of Vesalius and Harvey spanned just a few decades, the subsequent birth of scientific medicine was a long process, during which tradition remained alive and well. For example, up until the 1800s, many doctors still practiced Hippocratic medicine, including the use of laxatives, bleedings, and emetics to balance the humors. Nevertheless, two key figures stand out in the birth of modern scientific medicine: Ambroise Paré, whose pioneering work bridged the worlds of tradition and innovation; and René Laennec, who in 1816 invented a simple device that has been hailed one of the great discoveries in medicine—*and* a dark omen for the terrible turn Western medicine was about to take.

Ambroise Paré was a French military surgeon whose break with tradition in the mid-1500s has led many to call him the father of modern surgery. The title is justified given that Paré helped transform surgery—traditionally viewed as equivalent to butchery and reserved for barber surgeons with little training—into a professional art. But a closer look at his accomplishments shows that Paré had a healthy respect for innovation *and* tradition. Paré's best known discovery came in 1537 when he was working in the battlefields as a military surgeon and ran out of the oil that was traditionally used to treat gunshot wounds. At the time, gunshot wounds were thought to be poisonous and so were treated like snakebites, with boiling oil. With no oil at hand, Paré was forced to improvise and instead created an odd concoction of egg yolks, oil of roses, and turpentine. To his delight, the new formula was not only less painful for the soldiers, but also *more* effective. As he later wrote, "I resolved with myself never to so cruelly burn poor men wounded with gunshot." But along with such innovations, Paré had a healthy respect for tradition. In another milestone achievement, he revived the ancient art of ligature—tying off blood vessels to stop bleeding—in soldiers who required amputations, rather than cauterizing the wounds with red-hot irons. This, too, turned out to be a gentler and better way to stop bleeding and improve healing.

Paré's kinder approach to medicine—which also included making simple wooden legs for amputees who could not afford better devices—blended tradition and innovation, and eventually earned him the title, "The Gentle Surgeon." And in one final nod to tradition, Paré is famous for a humble comment he once made, echoing a similar point made by Hippocrates 2,000 years earlier: "I treated him, but God healed him."

If Paré's achievements in the 1500s represented a transitional phase when Western medicine overlapped two worlds, no milestone better symbolizes its shift into modernity than French physician René Laennec's milestone invention two centuries later: the stethoscope. At the time Laennec was practicing medicine, in the early 1800s, physicians typically listened to a patient's lungs and heart for signs of disease by placing their ears directly on the patient's chest or on a handkerchief laid on the chest. However, one day in 1816, Laennec was struggling with this method as he attempted to examine a young obese woman suffering from advanced heart disease. Frustrated by her size, modesty, or both, he was unable to place his ear on her chest to listen to her heart.

That's when inspiration struck. Laennec suddenly remembered a recent memory of two children he had seen playing in a park. At one point, they had picked up a long stick and, placing an end in their ears, began lightly tapping signals to each other. Recalling how the stick had amplified and transmitted the sounds, Laennec was suddenly struck by an idea. He quickly found a "quire" of paper (24 sheets) and rolled it into a cylinder. Then he placed one end in his ear, the other end on the woman's chest, and began to listen. Laennec later wrote that he "was not a little surprised and pleased to find that I could thereby perceive the action of the heart in a manner much more clear and distinct than I had ever been able to do by the immediate application of my ear."

Laennec went on to build sturdier versions of his stethoscope and used them to make many important discoveries, not only about how the new device could be used to amplify heart sounds, but how those sounds provided important clues about normal heart function and heart disease. Three years later, he published his findings in a landmark work, *On Mediate Auscultation or Treatise on the Diagnosis of the Diseases of the Lungs and Heart*, with the term "mediate auscultation" referring to "indirect listening." But despite his discovery and clinical observations,

Laennec's stethoscope was met with criticism and skepticism for decades. As late as 1885, one professor of medicine was famously quoted as stating, "He that has ears to hear, let him use his ears and not a stethoscope." Even Lewis A. Conner, founder of the American Heart Association, opted for placing his ear on a handkerchief over the chest, rather than Laennec's stethoscope.

But despite this resistance, the stethoscope was welcomed by many physicians and today is seen as a defining symbol in the birth of modern medicine, for good *and* bad. On the plus side, the stethoscope was one of the first effective technologies to advance the science of medicine. Indeed, it is still used today to gather diagnostically valuable information. On the other hand, the stethoscope represented a giant step away from tradition, from when physicians placed their ears over the hearts of their patients in an act that couldn't help convey a sense of intimacy and caring. Like no other innovation before or after, the stethoscope put a small, chilly barrier between physician and patient.

Beyond the invention of the stethoscope, the birth of modern medicine arose from many other advances over the next 150 years, as documented earlier in this book. But the advent of the stethoscope signaled a turning point in the physician-patient relationship, a shift in how physicians would care for their patients. When patients finally began to rebel against that shift, a range of traditional alternatives were available, not only those from ancient times, but some born only a century or two before.

Milestone #4 The birth of alternative medicine: a healing touch and a disdain for "heroic" medicine

One shouldn't feel too badly about the long history of scorn and disdain scientific medicine has heaped upon alternative medicine. Alternative medicine itself was born in part from the scorn and disdain *it* held for scientific medicine. The attitude was understandable given the state of scientific medicine in the early 1800s. As described earlier, scientific medicine was just one of many competing healthcare systems at the time, with few successes and little to offer. In fact, to practitioners of other healing systems, scientific medicine had much to offer *against* it.

With its harsh attempts to "save" patients with bleedings, toxic purgatives, and surgery that often ended in death from infection, scientific medicine was often sarcastically called "heroic" medicine. This contempt was perhaps best summarized by Samuel Hahnemann, the founder of homeopathy, who called "heroic" medicine a "non-healing art…that shortened the lives of ten times as many human beings as the most destructive wars and rendered millions of patients more diseased and wretched than they were originally."

Hahnemann's use of the term "non-healing" was telling because, like the ancient forms of traditional medicine, many alternative therapies that arose in the 1800s shared a belief in the mysterious forces of healing, as well as the value of natural treatments and the healer-patient relationship. This was at odds with scientific medicine's tendency to aggressively "attack" diseases with surgery and drugs. But among the many healing systems born in the nineteenth century, two in particular—homeopathy and chiropractic—illustrate how widely diverse they can be in treatment approach, while retaining some crucial similarities in their underlying philosophies.

Homeopathy

For many people, modern alternative medicine began in the late 1700s when Samuel Hahnemann discovered a new and almost completely counterintuitive theory of medicine. Hahnemann was a German physician and was translating a text on herbal medicine when he came upon a statement that struck him as odd: The author claimed that quinine, which is derived from Cinchona bark, was able to cure malaria because of its "bitterness." This made no sense to Hahnemann. So, taking the matter into his own hands, he began ingesting repeated doses of Cinchona to personally experience its effects. When he found the bark actually *caused* malaria-like symptoms, he had his milestone insight: Perhaps quinine's curative powers against malaria came not from its bitterness, but its ability to cause symptoms similar to the disease it was being used to treat. And if that was true, perhaps *other* medicines could be developed based on how closely *they* mimicked the symptoms of a given disease. After testing his theory with numerous substances in many volunteers, Hahnemann concluded his hypothesis was correct. He called it the "principle of similars," or "like cures like."

As Hahnemann continued his experiments and developed his theory of homeopathy—from the Greek *omoios* (similar) and *pathos* (feeling)—he incorporated two other central ideas. The first and most counterintuitive idea was that while homeopathic remedies, by definition, caused unwanted symptoms, their toxicity could be reduced by diluting them over and over until they caused no symptoms. And though the amount of substance that remained after so many dilutions was vanishingly small, their therapeutic power could be *increased* by a process he called "potenization"—shaking the solution between dilutions to extract the "vital" or "spirit-like" nature of the substance. Hahnemann's second major idea was that selection of a particular homeopathic therapy must be based on the *total* character of a person's symptoms, which therefore required a detailed understanding of the patient's history and personal traits.

From these concepts, one can see how homeopathic medicine shares some underlying values seen in ancient traditional medicines. First, the "vital" energy of the medicinal substances echoes the ancient view of vital energies in the human body and their interconnections with the outer world. Second, Hahnemann's emphasis on understanding the entire "character" of a patient's symptoms reflects the importance of the healer-patient relationship. For example, to determine a patient's symptom "picture," practitioners must spend considerable time interviewing patients to learn not only their symptoms, but how symptoms are affected by such factors as time of day, weather, season, mood, and behavior. Once this information is gathered, practitioners can select one or more homeopathic medicines, which today include more than 2,000 remedies. Finally, homeopathy is similar to traditional medicine because its treatments are derived from natural products (for example, plants, animals, and minerals) and involve tiny amounts.

Not surprisingly, scientific medicine opposed the theory of homeopathy from the start, dismissing the notion that such highly diluted substances could have any therapeutic effect and attributing any apparent benefits to the placebo effect. Nevertheless, despite a large body of contradictory evidence, in recent years a number of well-designed studies have suggested that homeopathic treatments may be effective for some conditions, including influenza, allergies, and childhood diarrhea. What's more, today, scientists continue to investigate molecular mechanisms that could explain how homeopathic remedies might work, including the idea that complex interactions between the substance and diluting solvent may create a molecular "memory" that gives the final solution its therapeutic effects. In any case, despite many battles with scientific medicine throughout the nineteenth century, homeopathy has survived more than 200 years, and in 2007 was one of the top ten alternative medicine therapies in the United States.

Chiropractic

Chiropractic medicine—the manipulation of misaligned joints to treat a variety of ailments—has a colorful history dating back to Hippocrates, whose treatments for curvature of the spine included tying patients to a ladder and dropping them from roof-height. While modern chiropractic medicine is considerably more sophisticated, it is certainly no less colorful: It was developed in the late 1890s by Daniel David (D. D.) Palmer, a former magnetic healer who had a sixth-grade education and who claimed that "ninety-five percent of all diseases are caused by displaced vertebrae."

Although Palmer had worked as a magnetic healer long before discovering chiropractic medicine in 1895, he eventually combined his gift for healing with a modern biological twist: He believed that the essential feature of all disease was inflammation and that he could heal patients by channeling his "vital" magnetic energy through his hands and into the site of inflammation. According to one homeopathic physician who had witnessed Palmer in action, "He heals the sick, the lame, and those paralyzed through the medium of his potent magnetic fingers placed upon the organ or organs diseased.... Dr. Palmer seeks out the diseased organ upon which the disease depends, and treats that organ." It was not until 1895 that Palmer adopted a new concept that would eventually lead to his discovery of chiropractic: He theorized that diseases arose when organs and tissues became displaced and rubbed against each other, with the resulting friction causing inflammation. From this theory, he developed the idea that by manually repositioning, or manipulating, displaced body structures back into their normal position, he could stop the friction, which "cooled down" the inflammation and cured the disease.

Palmer's milestone moment came on September 18, 1895, when he tried his new technique on a janitor in his building. Harvey Lillard had lost his hearing after twisting his back, and during the examination, Palmer noticed that one of Lillard's vertebrae was out of position. Palmer manipulated the bone to realign it, and both patient and physician were astonished by the result. "I was deaf for 17 years," Lillard wrote later, "and I expected to always remain so, for I had doctored a great deal without any benefit.... Dr. Palmer treated me on the spine [and] in two treatments I could hear quite well. That was eight months ago. My hearing remains good." Palmer became further convinced when, a short time later, he successfully treated a patient with heart trouble. After more experiments and refinement, Palmer called his new treatment "chiropractic" (from the Greek word for "done by hand"), and within a year he had opened a training school.

Although Palmer's original technique applied to any displaced tissues in the body, by 1903 he was focusing only on joints—particularly the spine—based on the so-called "foot on the hose" theory. According to that theory, if vertebrae became misaligned, they could pinch the roots of nerves exiting the spine, interrupt the transmission of nerve impulses to various organs, and thereby cause inflammation and disease. Thus spinal manipulation, realigning misplaced vertebrae and removing the pressure they put on nerves, could theoretically relieve inflammation and disease anywhere in the body depending on which body tissue the nerve supplied.

While many chiropractors today have evolved beyond Palmer's original "one-cause, one-cure" thesis, this view has long been a key tenet of chiropractic theory and a major reason why scientific medicine has opposed the profession. But although there is no scientific proof to the broad theory that impingement on nerves by misaligned vertebrae causes disease, in 1994 an evidence-based consensus panel convened by the U.S. Agency for Health Care Policy and Research *did* find that spinal manipulation is an effective treatment for back pain. With other studies showing benefits in other specific conditions, chiropractic has been increasingly accepted by mainstream medicine, and by 2007 it was the fourth most commonly used alternative therapy in the United States. This popularity undoubtedly stems not only from its effectiveness, but from its roots in tradition—from Palmer's belief that he was tapping into the body's "innate healing intelligence," to its realization of the ultimate in physician-patient care: hands-on healing.

* * *

Apart from homeopathy and chiropractic, many other forms of alternative medicine born in the 1800s are alive and well today, including naturopathic medicine (which focuses on the healing power of nature and natural treatments) and osteopathic medicine (which emphasizes natural treatments and manipulation of the musculoskeletal system and today is a conventional system equivalent to scientific medicine). Despite differences in technique, all shared traditional values that would never completely disappear or lose their appeal. Unfortunately, they also shared something else: a long and bitter battle with scientific medicine.

Milestone #5 Battle for the body: scientific medicine's blinding quest to conquer "quackery"

Scientific medicine has never been shy to dismiss if not denigrate any perceived threat to its values or power. This was never better illustrated than in 1842—decades *before* scientific medicine had achieved most of its greatest milestones—when Harvard Medical School's Oliver Wendell Holmes threw down the gauntlet in the looming battle with alternative medicine. Responding to Hahnemann's earlier criticism of scientific medicine, Holmes thundered that homeopathy was a "mingled mass of perverse ingenuity, of tinsel erudition, of imbecilic credulity, and artful misrepresentation." Despite the ironies of ignorance—scientific medicine would soon be similarly vicious in dismissing John Snow and Ignaz Semmelweis' evidence of germ theory (Chapters 2 and 3)—Holmes' concern was understandable: Among the hodge-podge of medical systems brewing at the time, scientific medicine was, as much as any other, fearing for its own survival.

The rise of modern medicine over the next two centuries is impressive, given that in 1800 there were only about 200 educated physicians in the entire United States. And though this number would increase to several thousand by 1830, most patients were still getting their healthcare from such "specialists" as botanical healers, inoculaters, midwives, bonesetters, and various shady purveyors of ill-defined tonics and nostrums. To bring some order to this medical mish-mash, practitioners were often grouped into one of three general categories. The "regulars" included scientific, or conventional, physicians who practiced orthodox medicine; "irregulars" included practitioners of unconventional or unorthodox medicine such as homeopathy; and the remaining jumble of swindlers and dreamers fell into the category of "quacks and charlatans." But such labels could not hide the disturbing fact that as late as the 1840s, essentially *any*one could call himself a doctor. It was this alarming reality that

finally prompted the "regulars" to marshal their forces in 18
achieve a milestone that would have an historic and long-lasting impact: At a national gathering of state medical societies in Philadelphia, they formed the National Medical Association—soon to be renamed the American Medical Association (AMA).

From its inception, the AMA had many noble goals, including raising the standards of medical education, eliminating poorly trained conventional physicians, and advancing medical knowledge. In addition, by the end of the 1800s, the AMA's legislative influence had led to all states requiring a license to practice medicine. And when in the early 1900s the AMA was concerned about medical education standards and quality, it commissioned an investigation. The resulting 1910 Flexnor report was so critical of medical education that the subsequent changes transformed medical schools and led to the standards that still apply today, including the requirement that students receive two years of basic science followed by two years of clinical training.

But beyond raising its own standards, the AMA was continually engaged in a battle against the irregulars, those practitioners of "alternative" medicine whose unscientific methods and philosophies jarred with their own. But though AMA efforts were largely successful—many fringe alternative systems were in decline by the late 1800s—it discovered to its dismay that one particularly grating system, homeopathy, was actually *gaining* followers. And so in 1876, the AMA initiated a strategy it would often use to squeeze out those whose values did not align with its own: It passed a resolution making it unethical to associate with homeopathic practitioners.

The AMA's battle and shunning strategies would only escalate in the twentieth century. As late as the 1950s, it still considered osteopathy to be quackery, and in the 1960s it created a "Committee on Quackery" against chiropractors, whose goals included "first the containment" and ultimately "the elimination" of chiropractic. But the AMA's cleansing tactics reached an all-time low in the 1960s, when it stooped to suppressing research favorable to chiropractic and initiated a campaign of misinformation that portrayed chiropractors as "unscientific, cultist, and having philosophy incompatible with western scientific medicine." Despite such tactics, by 1974 chiropractors had gained legal recognition in all states, and in 1987 the U.S. Supreme Court upheld a lower-court decision that found the AMA guilty of antitrust violations in its attempt to eliminate the chiropractic profession. Nevertheless, in a statement that summarized scientific medicine's bruising impact on many forms of conventional medicine, one federal judge concluded that "the injury to chiropractors' reputations… has not been repaired."

* * *

Despite the battles and the damage done, it would be foolish to underestimate the importance that rigorous standards of scientific medicine played in the breakthroughs that saved millions of lives throughout the nineteenth and twentieth centuries. What's more, these advances did not always come at the expense of patient caring. Up until the 1920s and 1930s, many conventional physicians still put patient care at the center of their practices. But by the 1930s and 1940s, a shift in priorities was clearly underway. Scientific medicine was becoming progressively removed from patients, just as it progressively removed *patients* from control over their own care. By the final decades of the twentieth century, patients sought to regain that control by reaching for new alternatives.

Milestone #6 Prescription for an epidemic (of frustration): the rediscovery of alternative medicine

The wake-up call finally arrived in a double-punch from the two most respected voices in scientific medicine. The first blow landed in 1993 when a study by David M. Eisenberg and associates in the *New England Journal of Medicine* found that, based on a 1990 national survey, 34%—more than *a third*—of respondents had used at least one "unconventional" therapy in the past year. Even more stunning: Projected to a national level, the study found that there were more visits to alternative practitioners than to *all* primary care physicians. The second punch arrived five years later in a 1998 follow-up study in the *Journal of the American Medical Association*: Since the first study, alternative medicine use had increased "substantially," with more than 42% of respondents reporting they had used at least one form of alternative therapy in 1997.

The turning point had arrived. With the evidence-based writing on its own wall, there was no turning back. The authors of the 1998 study, noting that "our survey confirms that alternative medicine use and expenditures have increased dramatically," concluded with a call for a "more proactive posture," more research, and the development of better educational curricula, credentialing, and referral guidelines. Scientific medicine had finally accepted a long-denied reality and was now actually inviting its black sheep relation into the family—*providing* it cleaned up its act.

As it turned out, the cleaning was already underway. In that same year (1998), Congress formally established the National Center for Complementary and Alternative Medicine (NCCAM), a landmark agency that bridged two long-separated worlds of medicine. As one of 27 institutes and centers within the National Institutes of Health (NIH), NCCAM's mission included exploring complementary and alternative medicine (CAM) "in the context of rigorous science." By 2009, it had an annual budget of $125.5 million—up from $19.5 million in 1998—and had funded more than 1,200 research projects throughout the world.

Since its inception, the NCCAM has gone a long way to help define, explain, legitimize, and sometimes debunk the many therapies found in the world of alternative medicine. For example, NCCAM broadly defines CAM as "a group of diverse medical and healthcare systems, practices, and products that are not generally considered to be part of conventional medicine." It also distinguishes between complementary medicine, treatments used *with* conventional medicine (such as aromatherapy to lesson discomfort after surgery); and alternative medicine, treatments used *instead of* conventional medicine (such as a special diet to treat cancer instead of radiation or chemotherapy).

And while it's difficult to classify and sort out the many different types of alternative medicine, the NCAAM groups CAM into four major categories: mind-body, biologically-based, manipulative/body-based, and energy. In addition, the broad category of "whole medical systems" includes those from both Western cultures (homeopathy and naturopathic medicine) and non-Western cultures (traditional Chinese medicine and Ayurvedic medicine). The NCCAM also provides information on specific therapies and recent findings. For example, in 2008, it released results from the National Health Interview Survey (NHIS), which showed that in 2007, *half* of all Americans—38% of adults and 12% of children—used some form of CAM. As shown in the accompanying figure, the study also found that the most five common CAM therapies were natural products, deep breathing, meditation, chiropractic and osteopathic treatment, and massage.

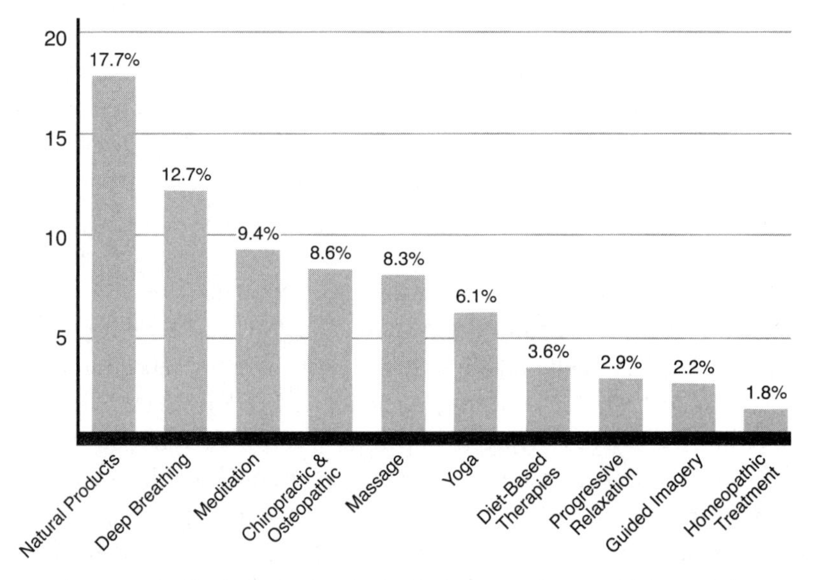

Source: CDC/NCHS, National Health Interview Survey, 2007 (NCCAM, 2008).

FIGURE 10.1 Top 10 alternative medicine therapies in 2007

The NHIS survey also revealed which CAM practitioners patients visit most. As shown in Table 10.1, the top two were chiropractic or osteopathic practitioners and massage therapists, accounting for more than 36 million visits.

TABLE 10.1 Top 15 CAM Practitioner Visits in 2007

Rank	Therapy	General Definition	Visits
1	Chiropractic/ Osteopathic	*Chiropractic*: Therapy based on relationship between body structure and function (usually spine) and effects on health. *Osteopathic*: Form of conventional medicine that emphasizes diseases arising in musculoskeletal system.	18,740,000
2	Massage	Manipulation of muscle and connective tissue to enhance function and promote relaxation and well-being.	18,068,000
3	Movement	Therapies focusing on movement of one or more parts of the body (including Feldenkreis, Alexander technique, and Pilates).	3,146,000

TABLE 10.1 Continued

Rank	Therapy	General Definition	Visits
4	Acupuncture	Form of traditional Chinese medicine (TCM); involves stimulating anatomical points on the body (such as by needles), which are manipulated by hands or electrical stimulation.	3,141,000
5	Relaxation techniques	Includes meditation, guided imagery, biofeedback, and deep breathing.	3,131,000
6	Natural products	Refers to non-mineral, non-vitamin natural products.	1,488,000
7	Energy healing	Therapies based on the use of energy fields; includes Reiki, qi gong, crystals, magnetic fields, and therapeutic touch.	1,216,000
8	Homeopathic treatment	Whole medical system in which patients are treated with highly diluted substances that in greater concentrations would cause symptoms of the condition being treated.	862,000
9	Traditional healers	Traditional healing systems developed within or outside Western cultures (for example, Curandero, Shaman, Native American).	812,000
10	Naturopathic treatment	Whole medical system that supports the body's natural healing power; incorporates various treatment approaches (for example, nutrition, lifestyle, medicinal plants, exercise).	729,000
11	Hypnosis	Mind-body therapy in which patients are induced into a trancelike state that resembles sleep and in which suggestions are readily accepted.	561,000
12	Biofeedback	Mind-body therapy in which involuntary bodily processes (such as heartbeat) are made perceptible so they can be altered by conscious mental control.	362,000
13	Diet-based therapies	Biologically based therapy that includes vegetarian, macrobiotic, Atkins, Pritikin, Ornish, and South Beach diets.	270,000

Rank	Therapy	General Definition	Visits
14	Ayurvedic medicine	Whole medical system developed in India; emphasizes connection among body, mind, and spirit, and includes diet and herbal treatments.	214,000
15	Chelation therapy	Biologically based therapy in which a chemical treatment is used to remove heavy metals from the body.	111,000
Total			**38,146,000***

*The total exceeds the sum of individual treatments because survey respondents could select more than one therapy.

But perhaps the most revealing findings from the 2007 NHIS study were the most common reasons *why* patients seek CAM practitioners. Among the top five reasons, all were *chronic* conditions: back pain (17.1%), neck pain (5.9%), joint pain (5.2%), arthritis (3.5%), and anxiety (2.8%). This underscores why the rediscovery of alternative medicine has emerged as one of the top ten breakthroughs in medicine. With Western medicine's focus on subspecialization and dividing the body into smaller and smaller pieces, it has often failed to successfully treat patients suffering from chronic conditions and pain affecting the *entire* body. For many patients, that need is often better met by alternative medicine whether because of its focus on holistic balance, more natural treatments, or a more traditional healer-patient relationship.

Listening: the unexpected phenomenon that triggered a transformation

Since the early 1800s, alternative and scientific medicine have struggled in a tug of war over philosophy, values, and methods—one side pulling patients in the direction of tradition, natural treatments, and a closer physician-patient relationship; the other tugging back with the allure of technology, tests, and harsh but effective treatments. But in the final decades of the twentieth century, just when the shouting seemed to be loudest, an unexpected phenomenon triggered a transformation: Each side began *listening* to the other. Many in the world of alternative medicine began to realize—as the AMA had understood 150 years earlier—that their credibility and success would depend on better research and higher educational and practice standards. For example, chiropractic now requires four years of training and has standardized examinations and licensing in every state, and a recent surge of research is looking more closely than ever before at its methods.

At the same time, scientific medicine has opened its ears and mind to the shift in patient attitudes and a new consumer-oriented healthcare system. Physicians have begun to accept that patients are demanding more power over their own healthcare decisions, including the use of alternative medicine when conventional treatments fail. Other factors underlying the transformation range from society's increasing acceptance of cultural, ethnic, and religious diversity to physicians' own frustration with how technology and other trends have diminished their relationships with patients.

And so, in a surprisingly short time, the practice of medicine has undergone a remarkable change impacting patients, physicians, and institutions. And, perhaps most important, many would argue that this transformation was not simply additive.

Greater than the sum of its parts: a new integrative medicine

The rediscovery of alternative medicine was more than a handshake extended through a barb-wired fence. From the start, many healthcare practitioners recognized an opportunity to create a new kind of medicine—an *integrative medicine*—that would unite the best of both worlds and thus transcend the shortcomings of each. Integrative medicine has been defined as a "healing oriented medicine that takes account of the whole person (body, mind, and spirit), including all aspects of lifestyle. It emphasizes the therapeutic relationship and makes use of all appropriate therapies, both conventional and alternative." This definition was provided by the pioneering Program in Integrative Medicine at the University of Arizona, which was started in 1997 by physician Andrew Weil. As the country's first fellowship program in integrated medicine, its goals include teaching physicians about "the science of health and healing" and "therapies that are not part of Western medical practice." Since that time, it has been joined by a number of other fellowship programs and a Consortium of Academic Health Centers for Integrative Medicine that includes more than 30 medical schools.

Despite its promise, integrative medicine faces many challenges, including holding alternative treatments to the same standards as those held by scientific medicine. While the answer may seem to lay in the mantra of evidence-based medicine—randomized placebo-controlled clinical trials that objectively study whether or not treatments work—conducting such studies can be problematic. For example, the nature of many alternative therapies—with individualized or experiential treatments whose benefits are difficult to measure—can make such testing difficult if not impossible. Still, many alternative therapies are now undergoing rigorous scientific testing through the funding efforts of the NCCAM and others. In the meantime, advocates of integrative medicine stress that their goal is to make use of all *appropriate* therapies in both scientific and alternative medicine, while addressing the shortcomings of each.

Successful partnerships between alternative and scientific medicine have already formed. For example, a 2008 article in *Current Oncology* described "integrative oncology" as the next step in the evolution of cancer care, noting that the goals include supporting cancer patients and their families by improving quality of life, relieving the symptoms caused by conventional treatments, and in some cases enhancing conventional treatment. As one example, the authors wrote that "after a careful review of the available evidence," the Society for Integrative Oncology now supports acupuncture as a complementary therapy when cancer-related pain is poorly controlled.

While the jury is still out on whether many alternative therapies can safely and effectively be used as alternatives or complements to scientific medicine, the textbook *Integrative Medicine* notes that an integrative approach offers many benefits, including removing barriers to the body's natural healing response; using less invasive interventions before costly invasive procedures; facilitating healing by engaging mind, body, spirit, and community; providing care based on "continuous healing relationships" rather than "visits"; and giving patients more control over their own treatments. As a 2001 editorial in the *British Medical Journal* (*BMJ*) concluded, "... integrated medicine is not just about teaching doctors to use herbs instead of drugs. It is about restoring core values that have been eroded by social and economic forces. Integrated medicine is good medicine... and its success will be signaled by dropping the adjective."

A better day for Eastern medicine: balance regained

> "It is much more important to know what sort of patient has a disease, than what sort of disease a patient has."
>
> —*Sir William Osler (1849–1919); Canadian physician, father of modern medicine*

You'll recall that shortly after Yonten and his fellow Buddhist monks arrived in the United States, the memories of their traumatic experiences in Tibet were causing various symptoms and interfering with their ability to meditate. Although traditional Tibetan healers had diagnosed the problem as a life-wind imbalance, once in the U.S., the monks were referred to the Boston Center for Refugee Health and Human Rights for additional help. Psychiatrists at the Center had no objection to the Tibetan healers' diagnosis of *srog-rLung* but added a diagnosis of their own: post-traumatic stress disorder (PTSD). Then, using the principles of integrative medicine, clinicians worked with the monks to develop a treatment that combined both traditional *and* conventional medicine. The monks are now doing better thanks to an integrative—and more balanced—approach that includes not only breathing exercises, herbs, mantras, and singing bowls, but *also* a Western regimen of psychotherapy and antidepressant drugs.

* * *

The breakthrough rediscovery of alternative medicine and the emergence of integrative medicine is intuitively appealing because it embodies the best of what medicine has learned over thousands of years, from its culturally diverse origins in China, India, and Greece, to its revolutionary break with tradition during the Renaissance; from the rolled up paper in 1816 that led to the first stethoscope, to nearly two centuries of animosity between scientific and alternative medicine. Today, many believe that medicine can best achieve its potential through life-saving technologies and drugs, *and* traditional values that respect mind, body, spirit, and the physician-patient relationship. Since the time of Hippocrates and a millennium before, healers have known that curing isn't always possible. But caring—and therefore, *healing*—is.

Epilogue

"The greatest obstacle to discovery is not ignorance; it is the illusion of knowledge."

—Daniel J. Boorstin (American author and historian)

"A wild idea..." that is "at variance with established knowledge."

—The English Royal Society, rejecting Edward Jenner's discovery of the first vaccine

As the ten greatest breakthroughs in medicine have repeatedly shown, knowledge is a risky business. It can blind us from the very discoveries we seek. It can be so corroded by tradition that it blinds *others*. And, despite how much we try to accumulate, it can be overmatched by simple dumb luck. Somehow the individuals in this book navigated these challenges to arrive at ten breakthrough discoveries: 1) medicine itself; 2) sanitation; 3) germ theory; 4) anesthesia; 5) X-rays; 6) vaccines; 7) antibiotics; 8) genetics and DNA; 9) drugs for mental illness; and 10) alternative medicine. Looking back on their journeys, four lessons stand out as guideposts and advice for those seeking the next great breakthrough.

Lesson 1: Pay attention to the peculiar—and the obvious.

In the early 1800s, René Laennec was walking through a park when he noticed two children tapping signals to each other through a long stick held up to their ears. A short time later, while struggling to listen to the heart of an obese woman, his sudden recollection of that curious memory inspired him to roll up a tube of paper and invent the stethoscope, a milestone event that influenced the development of modern medicine. (Chapter 10)

In the early 1830s, when John Snow was sent to help miners who had been stricken by an outbreak of cholera, he noticed two odd things: 1) The workers were so deep underground that they *couldn't* be exposed to the "miasmatic" vapors thought to cause the disease, and 2) The miners ate their meals in close quarters to where they relieved their bowels. Fifteen years later, both observations helped inspire his revolutionary theory that cholera was transmitted by contaminated water, a key breakthrough in the discovery of germ theory. (Chapter 2)

In 1910, biologist Thomas Hunt Morgan found it peculiar when, after having bred millions of fruit flies with red eyes, he found a fly that had been born with *white* eyes. Following up on this peculiar finding ultimately led Morgan and his students to the milestone discovery that the basic units of heredity—genes—are located on chromosomes. (Chapter 8)

Lesson 2: Stick to your convictions despite doubt and ridicule.

In the late 1700s, Edward Jenner discovered that people could be protected from deadly smallpox infections by inoculating them with the far less dangerous *cow*pox. Despite a storm of protest from those who objected on scientific, religious, and moral grounds, Jenner persisted with his investigations. Within several years, his vaccine was saving countless lives throughout the world. (Chapter 6)

When Ignaz Semmelweis theorized in 1847 that a deadly infection was being spread through his hospital by physicians' unclean hands, he instituted a handwashing procedure that subsequently saved countless lives. Although the medical community ridiculed his belief that handwashing could stop disease transmission, Semmelweis refused to back down and is now credited with playing a major role in the discovery of germ theory. (Chapter 3)

In 1865, after ten years of experiments and growing thousands of pea plants, Gregor Mendel discovered the new field—and the first laws—of genetics. Although biologists ignored or downplayed his findings for the next three decades, Mendel contended until he died that "The time will come when the validity of the laws…will be recognized." Mendel was right; today, he is recognized as the father of genetics. (Chapter 8)

Lesson 3: Have the good sense to embrace your good luck.

In 1928, Alexander Fleming returned to his laboratory after a long vacation and found that one of his experiments had been ruined by a mold that he found growing in one of his bacterial cultures. Fleming turned this odd bit of fortune—and several other coincidences he wasn't even aware of—to his advantage and subsequently discovered penicillin, the first antibiotic. (Chapter 7)

In 1948, John Cade was studying patients with manic-depression, hoping he might find a toxic substance in their urine that would explain their bizarre behavior. But instead of finding a substance that caused mania, he stumbled upon a chemical that *prevented* it. Pursuing this unexpected turn of fate, Cade developed lithium carbonate, the first effective drug for mania. (Chapter 9)

In the early 1950s, James Watson and Francis Crick were among the many scientists struggling to figure out the structure of DNA. Then, in early 1953, Crick had the good fortune of being shown an X-ray image of DNA made by a competitor scientist, Rosalind Franklin. A short time later, the image sparked a flash of insight that helped Crick solve the mystery of DNA. (Chapter 8)

Lesson 4: Be willing to face down blind authority and tradition.

It's unclear how much flack Hippocrates took when he claimed in 400 BC that disease is not caused by evil spirits, but natural factors. Nevertheless, his courageous assertion broke with a cultural belief in superstition that dated back at least 600 years. Thanks to this and other milestone insights, Hippocrates is now credited with the discovery of medicine. (Chapter 1)

During the Renaissance of the sixteenth and seventeenth centuries, the work of two individuals—Andreas Vesalius with his stunning revelations of human anatomy and William Harvey with his landmark discovery of how blood circulates—gave birth to a new world of scientific medicine. But just as remarkable was their willingness to contradict the ancient teachings of Galen, whose authority had gone unquestioned for more than *1,200* years. (Chapter 10)

Traditional and alternative medicine have long advocated healing methods that focus on nontoxic therapies, restoring inner balance, and the healer-patient relationship. After standing up to two centuries of criticism and suppression by Western scientific medicine, alternative medicine re-emerged in the late 1990s by overwhelming patient demand. Today, this breakthrough is part of a new "integrative" and holistic medicine that offers the best of *both* worlds. (Chapter 10)

The H1N1 outbreak: lessons learned?

In the spring of 2009, a highly contagious epidemic broke out and spread like wildfire across the globe. No, not the H1N1 (swine) flu virus, but the *behavioral* epidemic that followed in its wake. Consider some of the social changes that resulted as an epidemic of fear swept across the world....

One of the first signs of the outbreak was the disappearance of hand sanitizer bottles from store shelves and their *re*appearance in purses, pockets, children's backpacks, and gym bags. By summer, the transportation industry had shifted into gear, as bus drivers and airline crews were trained to not only wipe down all surfaces with disinfectants, but to ask passengers how they felt and, if a fever was suspected, wave them goodbye. By fall, the world of religion had seen the light and was adopting new rituals, from Catholic priests withholding the public sharing of consecrated wine, to the installation of electronic dispensers that released holy water like so many squirts of soap in a public restroom. By November and December, Happy Holidays had become the Nightmare Before Christmas, as families opted for 10-hour drives in the hermetic safety of their cars rather than 90 minutes of incubation in an airplane cabin. Mall Santas, meanwhile, were tucking bottles of Purell behind the fuzzy cuffs of their shiny, black boots. By the end of 2009, other sightings of the ongoing behavioral epidemic included new ways to sneeze (into your bent elbow) and congratulate opponents after sports competitions (elbow bumps instead of hand shakes).

And so the question: What did the global response to H1N1 tell us about how much the world has changed since the breakthrough discoveries of sanitation, germ theory, vaccines, and others? It's easy to argue that little has changed since Ignaz Semmelweis demonstrated 160 years ago that handwashing can reduce hospital infections and deaths. For example, although nearly 100,000 people in the United States die each year from hospital-acquired infections, about one out of every two physicians *still* fails to follow recommended hand-washing guidelines. Public adherence is even worse, with one study showing that only 34% of men washed their hands after using the toilet. At the same time, many people today still regard vaccines with a feeling of queasy ambivalence. As some physicians have noted, the H1N1 epidemic showed how public reaction can go through cycles of fear: When H1N1 first broke out, people stormed their physicians' offices clamoring for a vaccine, fearful that the next Bubonic plague was underway. Six months later, after the initial panic died down, many of the same people ran the other way, fearful that the H1N1 vaccine might be harmful, despite many reports documenting its safety.

Nevertheless.

There's little doubt that the greatest breakthroughs in medicine have saved countless lives *and* changed our view of the world. While ignorance, carelessness, and irrational fear may be endemic to the human condition, every medical advance helps inoculate us a little better against our own worst natures. If that were not true, many of the responses against H1N1—from Purell bottles in Santa's boots to automated holy water dispensers—would not have occurred in the first place. It's impossible to know how deadly H1N1 or any other recent epidemic would have been without such transformations, but it seems safe to assume that the toll would have been much worse.

Perhaps one take-home lesson common to all ten breakthroughs is that regardless of where in the body the latest discovery impacts health and disease, its success will largely be determined by its influence on one area in particular—the human mind.

Appendix A

The Milestones, by Discovery

Chapter 1: Hippocrates and the Discovery of Medicine
Milestone #1: Getting real: diseases have natural causes
Milestone #2: It's the patient, stupid: the creation of clinical medicine
Milestone #3: A code of ethics that stands the test of time
Milestone #4: Acting the part: professionalizing the practice of medicine
Milestone #5: The enigmatic Corpus: 60 books and a wealth of medical firsts
Milestone #6: Where the two worlds meet: a holistic approach to medicine
Chapter 2: The Discovery of Sanitation
Milestone #1: The first epidemic: a lesson from the depths of a coal mine
Milestone #2: Casting aside miasma to envision a new kind of killer
Milestone #3: The invention of epidemiology and the disabling of a deadly pump
Milestone #4: A new "Poor Law" raises hackles—and awareness
Milestone #5: A grand report creates a wealth of ideas and a will to act
Milestone #6: The long, slow birth of a public health revolution
Chapter 3: The Discovery of Germs and How They Cause Disease
Milestone #1: The tragic loss of a friend (and a brilliant gain of insight)
Milestone #2: A simple solution: wash your hands and save a life
Milestone #3: From fermentation to pasteurization: the germination of germ theory
Milestone #4: The "spontaneous generation of life" finally meets its death
Milestone #5: The critical link: germs in the world of insects, animals, and people
Milestone #6: Antiseptics to the rescue: Joseph Lister and the modern age of surgery

Chapter 8: The Discovery of Heredity, Genetics, and DNA
Milestone #1: From peas to principles: Gregor Mendel discovers the rules of heredity
Milestone #2: Setting the stage: a deep dive into the secrets of the cell
Milestone #3: The discovery—and dismissal—of DNA
Milestone #4: Born again: resurrection of a monastery priest and his science of heredity
Milestone #5: The first genetic disease: kissin' cousins, black pee, and a familiar ratio
Milestone #6: Like beads on a necklace: the link between genes and chromosomes
Milestone #7: A transformational truth: the rediscovery of DNA and its peculiar properties
Milestone #8: Like a child's toy: the secrets of DNA and heredity finally revealed
Milestone #9: The great recount: human beings have *how* many chromosomes?
Milestone #10: Breaking the code: from letters and words to a literature of life

Chapter 9: The Discovery of Drugs for Madness, Sadness, and Fear
Milestone #1: Sickness, seizure, surgery, and shock: the first medical treatments for mental illness
Milestone #2: Mastering mania: lithium takes on the "worst patient on the ward"
Milestone #3: Silencing psychosis: chlorpromazine transforms patients *and* psychiatry
Milestone #4: Recovering the ability to laugh: the discovery of antidepressants
Milestone #5: More than "mother's little helper": a safer and better way to treat anxiety

Chapter 10: The Rediscovery of Alternative Medicine
Milestone #1: The birth of traditional medicine: when caring was curing
Milestone #2: Enlightenment: 1,200 years of tradition overturned and a new course for medicine
Milestone #3: The birth of scientific medicine: when curing began to overshadow caring
Milestone #4: The birth of alternative medicine: a healing touch and a disdain for "heroic" medicine
Milestone #5: Battle for the body: scientific medicine's blinding quest to conquer "quackery"
Milestone #6: Prescription for an epidemic (of frustration): the rediscovery of alternative medicine

Appendix B

References and Additional Reading

Chapter 1

Anaxagoras Fragments and Commentary. Fairbanks A., ed. and trans. *The First Philosophers of Greece*. London, UK: Paul, Trench, Trubner. 1898:235–262 from Hanover Historical Texts Project, http://history.hanover.edu/texts/presoc/anaxagor.htm.

Bujalkova, M. 2001. Hippocrates and his principles of medical ethics. *Bratislavské lekárske listy* 102(2):117–120.

Burnham, John. 2005. *What Is Medical History?* Cambridge, UK: Polity Press.

Chang A., E.M. Lad, and S.P. Lad. 2007. Hippocrates' influence on the origins of neurosurgery. *Neurosurgical Focus* 23(1) (July): 1–3.

Conrad, Lawrence I., Michael Neve, Vivian Nutton, Roy Porter, and Andrew Wear. 1995. *The Western Medical Tradition: 800 BC to AD 1800*. New York: Cambridge University Press.

Katsambas, A. and S.G. Marketos. 2007. Hippocratic messages for modern medicine (the vindication of Hippocrates). *Journal of the European Academy of Dermatology and Venereology* 21:859–861.

Longrigg, James. 1998. *Greek Medicine: From the Heroic to the Hellenistic Age*. London: Gerald Duckworth & Co. Ltd.

Nutton, Vivian. 2004. *Ancient Medicine*. London: Routledge.

Orfanos, C.E. 2007. From Hippocrates to modern medicine. *Journal of the European Academy of Dermatology and Venereology* 21:852–858.

Simopoulos, A.P. 2001.The Hippocratic concept of positive health in the 5th century BC and in the new millennium. Simopoulos, A.P. and K.N. Pavlou, ed. *Nutrition and Fitness: Diet, Genes, Physical Activity and Health*. Basel: Karger, 89:1–4.

Taylor, Henry Osborn. 1922. *Greek Biology and Medicine*. Boston: Marshall Jones Company, www.ancientlibrary.com/medicine/0002.html.

U.S. National Library of Medicine. 2002. *Greek Medicine*. National Institutes of Health, www.nlm.hih.gov/hmd/greek/index.html.

Chapter 2

Bentivoglio, M. and P. Pacini. 1995. Filippo Pacini: A determined observer. *Brain Research Bulletin* 38(2):161–165.

Cameron, D. and I.G. Jones. 1983. John Snow, the Broad Street pump, and modern epidemiology. *International Journal of Epidemiology* 12:393–396.

The Encyclopædia Britannica: A Dictionary of Arts, Sciences, Literature, and General Information. "Cholera." 11th ed. Volume VI. 1910. New York: The Encyclopædia Britannica Company.

Eyler, J.M. 2004. The changing assessments of John Snow's and William Farr's cholera studies. A. Morabia, ed. *A History of Epidemiologic Methods and Concepts*. Basel, Switzerland: Birkhäuser Verlag.

Halliday, S. 2001. Death and miasma in Victorian London: An obstinate belief. *British Medical Journal* 323 (December): 1469–1471.

Hamlin, C. and S. Sheard. 1998. Revolutions in public health: 1848 and 1998. *British Medical Journal* 317 (August): 587–591.

Howard-Jones, N. 1984. Robert Koch and the cholera vibrio: A centenary. *British Medical Journal* 288 (February): 379–381.

The Medical Times and Gazette: A Journal of Medical Science, Literature, Criticism, and News. 1853. The Cholera, Volume 7. London: John Churchill.

Melosi, Martin. 2000. *The Sanitary City: Urban Infrastructure in America from Colonial Times to the Present*. Baltimore: The Johns Hopkins University Press.

Newsom, S.W.B. 2006. Pioneers in infection control: John Snow, Henry Whitehead, the Broad Street pump, and the beginnings of geographical epidemiology. *Journal of Hospital Infection* 64:210–216.

Paneth, N. 2004. Assessing the contributions of John Snow to epidemiology 150 years after removal of the broad street pump handle. *Epidemiology* September 15(5):514–516.

Peterson, J.A. 1979. The impact of sanitary reform upon American urban planning, 1840–1890. *Journal of Social History* 13(1):83–103.

Reidl, J. and K.E. Klose. 2002.Vibrio cholerae and cholera: Out of the water and into the host. *FEMS Microbiology Reviews* 26:125–139.

Sack, D.A., R.G. Sack, G. Nair, and A.D. Siddique. 2004. Cholera. *The Lancet* 363 (January): 223–233.

Sellers, D. 1997. *Hidden Beneath Our Feet: The Story of Sewerage in Leeds*. Leeds: Leeds City Council, Department of Highways and Transportation (October).

Smith, C.E. 1982. The Broad Street pump revisited. *International Journal of Epidemiology* 11:99–100.

Snow, J. 1855. *Mode of Communication of Cholera*. London: John Churchill, www.ph.ucla.edu/epi/snow.html.

Snow, John website. Department of Epidemiology, School of Public Health, University of California, Los Angeles, www.ph.ucla.edu/epi/ snow.html.

Winterton, W.R. 1980. The Soho cholera epidemic 1854. *History of Medicine* (March/April): 11–20.

Zuckerman, J.N., L. Rombo, and A. Fisch. 2007. The true burden and risk of cholera: Implications for prevention and control. *The Lancet Infectious Diseases* 7:521–30.

Chapter 3

Bardell, D. 1982. The roles of sense and taste and clean teeth in the discovery of bacteria by Antoni van Leeuwenhoek. *Microbiological Reviews* 47(1) (March): 121–126.

Barnett, J.A. 2003. Beginnings of microbiology and biochemistry: The contribution of yeast research. *Microbiology* 149:557–567.

Baxter, A.G. 2001. Louis Pasteur's beer of revenge. *Nature Reviews Immunology* 1 (December): 229–232.

Blaser, M.J. 2006. Who are we? Indigenous microbes and the ecology of human diseases. *European Molecular Biology Organization (EMBO) Reports* 7(10):956–960.

Carter, K. Codell. 1981. Semmelweis and his predecessors. *Medical History* 25:57–72.

Carter, K. Codell. 1985. Koch's postulates in relation to the work of Jacob Henle and Edwin Klebs. *Medical History* 29:353–374.

Centers for Disease Control and Prevention. *Estimates of Healthcare–Associated Infections*. www.cdc.gov/ncidod/dhqp/hai.html.

Centers for Disease Control and Prevention. Guideline for Hand Hygiene in Health-Care Settings. 2002. *Morbidity and Mortality Weekly Report* 51(RR-16) (October): 1–33.

Dunlop, D.R. 1927. The life and work of Louis Pasteur. *The Canadian Medical Association Journal* (November): 297–303.

Dunn, P.M. 2005. Ignac Semmelweis (1818–1865) of Budapest and the prevention of puerperal fever. *Archives of Disease in Childhood. Fetal and Neonatal Edition* 90:F345–F348.

Elek, S.D. 1966. Semmelweis and the Oath of Hippocrates. *Proceedings of the Royal Society of Medicine* 59(4) (April): 346–352.

The Encyclopædia Britannica: A Dictionary of Arts, Sciences, Literature, and General Information. 11th ed. Volume XVI. 1911. "Lister, Joseph." New York: The Encyclopædia Britannica Company.

Fleming, A. 1947. Louis Pasteur. *British Medical Journal* (April): 517–522.

Fleming, J.B. 1966. Puerperal fever: The historical development of its treatment. *Proceedings of the Royal Society of Medicine* 59 (April): 341–345.

Fred, E.B. 1933. Antony van Leeuwenhoek: On the three-hundredth anniversary of his birth. *Journal of Bacteriology* 25(1):1–18.

Godwin, William. 1928. *Memoirs of Mary Wollstonecraft*. London: Constable and Co., http://dwardmac.pitzer.edu/anarchist_archives/godwin/memoirs/toc.html.

Goldmann, D. 2006. System failure versus personal accountability—The case for clean hands. *The New England Journal of Medicine* 355(2) (July): 121–123.

Gordon, J.I., R.E. Ley, R. Wilson, et al. *Extending our view of self: The human gut microbiome initiative (HGMI)*, http://genome.gov/Pages/Research/Sequencing/SeqProposals/HGMISeq.pdf.

Kaufmann, S.H.E. 2005. Robert Koch, the Nobel Prize, and the ongoing threat of tuberculosis. *The New England Journal of Medicine* 353(23) (December): 2423–2426.

Kaufmann, S.H.E. and U.E. Schaible. 2005. 100th anniversary of Robert Koch's Nobel Prize for the discovery of the tubercle bacillus. *Trends in Microbiology* 13(10) (October): 469–475.

Klevens, R.M., J.R. Edwards, C.S. Richards, Jr., et al. 2007. Estimating health care-associated infections and deaths in U.S. hospitals, 2002. *Public Health Reports* 122 (March–April): 160–166.

Krasner, R.I. 1995. Pasteur: High priest of microbiology. *ASM News* 61(11): 575–578.

Louden, Irvine. *The Tragedy of Childbed Fever*. Oxford University Press, www.oup.co.uk/pdf/0-19-820499-X.pdf.

National Institutes of Health. *NIH Launches Human Microbiome Project*. December 19, 2007, www.nih.gov/news/pr/dec2007/od-19.htm.

Nobelprize.org. *Robert Koch: The Nobel Prize in Physiology or Medicine, 1905*, http://nobelprize.org/nobel_prizes/medicine/laureates/1905/ koch-bio.html.

Nuland, Sherwin B. 1979. The enigma of Semmelweis—an interpretation. *Journal of the History of Medicine* (July): 255–272.

Pasteur, Louis. On Spontaneous Generation. Address delivered at the Sorbonne Scientific Soiree, April 7, 1864. *Revue des cours scientifics* 1 (April 23, 1864): 257–264.

Porter, J.R. 1961. Louis Pasteur: Achievements and disappointments, 1861. *Pasteur Award Lecture* 25:389–403.

Porter, J.R. 1976. Antony van Leeuwenhoek: Tercentenary of his discovery of bacteria. *Bacteriological Reviews* 40(2) (June): 260–269.

Semmelweis, Ignaz. 1983. *The Etiology, Concept, and Prophylaxis of Childbed Fever*. Trans. K. Codell Carter. Madison, WI: The University of Wisconsin Press.

Tomes, N.J. 1997. American attitudes toward the germ theory of disease: Phyllis Allen Richmond revisited. *Journal of the History of Medicine* 52 (January): 17–50.

Ullmann, A. 2007. Pasteur-Koch: Distinctive ways of thinking about infectious diseases. *Microbe* 2(8):383–387.

Chapter 4

Adams, A.K. 1996. The delayed arrival: From Davy (1800) to Morton (1846). *Journal of the Royal Society of Medicine* 89 (February): 96P–100P.

Bigelow, H.J. 1846. Insensibility during surgical operations produced by inhalation. *Boston Medical and Surgical Journal* 35:309–317.

Burney, Fanny. *Eyewitness: Major Surgery Without an Anaesthetic, 1811*. Letters and journals of Fanny Burney, www.mytimemachine.co.uk/operation.htm.

Caton, Donald. 1999. *What a Blessing She Had Chloroform: The Medical and Social Response to the Pain of Childbirth From 1800 to the Present*. New Haven, CT: Yale University Press.

Clark, R.B. 1997. Fanny Longfellow and Nathan Keep. *American Society of Anesthesiologists Newsletter* 61(9) (September): 1–3.

Collins, Vincent J. 1993. *Principles of Anesthesiology*, 3rd edition. Philadelphia: Lea & Febiger.

Davy, Humphry. 1800. *Researches, Chemical and Philosophical; Chiefly Concerning Nitrous Oxide or Dephlogisticated Nitrous Air, and Its Respiration*. Bristol: Biggs and Cottle.

Desai, S.P., M.S. Desai, and C.S. Pandav. 2007. The discovery of modern anaesthesia—Contributions of Davy, Clarke, Long, Wells, and Morton. *Indian Journal of Anaesthesia* 51(6):472–476.

Greene, N.M. 1971. A consideration of factors in the discovery of anesthesia and their effects on its development. *Anesthesiology* 35(5) (November): 515–522.

Jacob, M.C. and M.J. Sauter. 2002. Why did Humphry Davy and associates not pursue the pain-alleviating effects of nitrous oxide? *Journal of the History of Medicine* 57 (April): 161–176.

Larson, M.D. 2005. History of Anesthetic Practice. Miller, R.D., ed. *Miller's Anesthesia*, 6th ed. Philadelphia: Elsevier Churchill Livingstone.

Morgan, G.E., M.S. Mikhail, M.J. Murray, eds. 2002. *Clinical Anesthesiology*, 3rd edition. New York: McGraw-Hill Professional.

Orser, B.A. 2007. Lifting the fog around anesthesia. *Scientific American* (June): 54–61.

Rudolph, U. and B. Antkowiak. 2004. Molecular and neuronal substrates for general anaesthetics. *Nature Reviews. Neuroscience* 5 (September): 709–720.

Smith, W.D.A. 1965. A history of nitrous oxide and oxygen anaesthesia, Part I: Joseph Priestley to Humphry Davy. *British Journal of Anaesthesia* 37:790–798.

Snow, John. 1847. *On the Inhalation of the Vapour of Ether in Surgical Operations: Containing a Description of the Various States of Etherization*. London: John Churchill.

Terrell, R.C. 2008. The invention and development of enflurane, isoflurane, sevoflurane, and desflurane. *Anesthesiology* 108:531–533.

Thatcher, Virginia S. 1984. *History of Anesthesia, With Emphasis on the Nurse Specialist*. New York: Garland Publishing, Inc.

Thoreau, H.D. *This Date, from Henry David Thoreau's Journal: 1851*, http://hdt.typepad.com/henrys_blog/1851/index.html.

Thornton, J.L. *John Snow, Pioneer Specialist-Anaesthetist*. John Snow website. Department of Epidemiology, School of Public Health, University of California, Los Angeles, www.ph.ucla.edu/epi/snow/anaesthesia5(5)_129_135_1950.pdf.

Trevor, A.J. and P.F. White. 2004. General anesthetics. *Basic & Clinical Pharmacology*, 9th ed. Katzung, B.G., ed. New York: Lange Medical Books/McGraw-Hill.

Chapter 5

American Institute of Physics website. *Marie Curie and the Science of Radioactivity*, www.aip.org/history/curie/war1.htm.

Assmus, A. 1995. Early History of X-Rays. *BeamLine* 25(2) (Summer): 10–24.

Bowers, Brian. 1970. *X-rays: Their Discovery and Applications*. London: Her Majesty's Stationery Office.

Brecher, Ruth and Edward Brecher. 1969. *The Rays: A History of Radiology in the United States and Canada*. Baltimore: The Williams and Wilkins Company.

Centers for Disease Control and Prevention. 2007. *Mammography*. National Center for Health Statistics (Health, United States, Table 87), www.cdc.gov/nchs/fastats/mammogram.htm.

Daniel, T.M. 2006. Wilhelm Conrad Röntgen and the advent of thoracic radiology. *The International Journal of Tuberculosis and Lung Disease* 10(11):1212–1214.

Doris, C.I. 1995. Diagnostic imaging at its centennial: The past, the present, and the future. *Canadian Medical Association Journal* 153(9) (November): 1297–1300.

Frame, P. *Coolidge X-ray Tubes*, www.orau.org/PTP/collection/xraytubescoolidge/coolidgeinformation.htm.

Frankel, R.I. 1996. Centennial of Röntgen's discovery of X-rays. *Western Journal of Medicine* 164:497–501.

Glasser, Otto. 1934. *Wilhelm Conrad Röntgen and the Early History of the Roentgen Rays*. Springfield, IL: Charles C. Thomas.

Hessenbruch, A. 1995. X-rays for medical use. *Physics Education* 30(6) (November): 347–355.

Kogelnik, H.D. 1997. Inauguration of radiotherapy as a new scientific specialty by Leopold Freund 100 years ago. *Radiotherapy and Oncology* 42:203–211.

Lentle, B. and J. Aldrich. 1997. Radiological sciences, past and present. *The Lancet* 350 (July): 280–85.

Linton, O.W. 1995. Medical applications of X Rays. *BeamLine* 25(2) (Summer): 25–34.

Mettler, Fred A., Jr. 2005. *Essentials of Radiology*, 2nd ed. Philadelphia: Elsevier Saunders.

Mould, R.F. 1995. The early history of X-ray diagnosis with emphasis on the contributions of physics, 1895–1915. *Physics in Medicine and Biology* 40:1741–1787.

New York Times. 1921. Dangers of x-ray: new investigation, following recent deaths, to insure scientists' protection. May 15.

Nobelprize.org. *Allan M. Cormack: The Nobel Prize in Physiology or Medicine, 1979*, http://nobelprize.org/nobel_prizes/medicine/laureates/1979/cormack-autobio.html.

Nobelprize.org. *Godfrey N. Hounsfield: The Nobel Prize in Physiology or Medicine, 1979*, http://nobelprize.org/nobel_prizes/medicine/laureates/1979/hounsfield-autobio.html.

Nobelprize.org. *Max von Laue: The Nobel Prize in Physics, 1914*, http://nobelprize.org/nobel_prizes/physics/laureates/1914/laue-bio.html.

Nobelprize.org. *Wilhelm Conrad Röntgen: The Nobel Prize in Physics, 1901*, http://nobelprize.org/nobel_prizes/physics/laureates/1901/rontgen-bio.html.

Posner, E. 1970. Reception of Röntgen's discovery in Britain and U.S.A. *British Medical Journal* 4 (November): 357–360.

Roentgen, W.C. 1896. On a new kind of rays. *Nature* 53:274–277.

Schedel, A. 1995. An unprecedented sensation—Public reaction to the discovery of x-rays. *Physics Education* 30(6) (November): 342–347.

Suits, C.G. *William David Coolidge: Inventor, Physicist, Research Director*, www.harvardsquarelibrary.org/unitarians/coolidge.html.

Sumner, D. 1995. X-rays—Risks versus benefits. *Physics Education* 30(6) (November): 338–342.

Wesolowski, J.R. and M.H. Lev. 2005. CT: History, Technology, and Clinical Aspects. *Seminars in Ultrasound, CT, and MRI* 26:376–379.

Chapter 6

André, F.E. 2001. The future of vaccines, immunization concepts and practice. *Vaccine* 19:2206–2209.

André, F.E. 2003. Vaccinology: Past achievements, present roadblocks, and future promises. *Vaccine* 21:593–595.

Atkinson, W., J. Hamborsky, L. McIntyre, and C. Wolfe, eds. 2008. *Epidemiology and Prevention of Vaccine-Preventable Diseases*, 10th ed. Centers for Disease Control and Prevention (February).

Barquet, N. and D. Pere. 1997. Smallpox: The triumph over the most terrible of the ministers of death. *Annals of Internal Medicine* 127:635–642.

Baxter, D. 2007. Active and passive immunity, vaccine types, excipients and licensing. *Occupational Medicine* 57:552–556.

Bazin, Hervé. 2000. *The Eradication of Smallpox*. San Diego: Academic Press.

Bazin, Hervé. 2003. A brief history of the prevention of infectious diseases by immunizations. *Comparative Immunology, Microbiology & Infectious Diseases* 26:293–308.

Behbehani, A.M. 1983. The smallpox story: Life and death of an old disease. *Microbiological Reviews* 47 (December): 455–509.

Broome, C.V. 1998. *Testimony on eradication of infectious diseases*. Delivered to the U.S. House Committee on International Relations. May 20, www.hhs.gov/asl/testify/t980520a.html.

Centers for Disease Control and Prevention. *Smallpox Overview*, www.bt.cdc.gov/agent/smallpox/overview/disease-facts.asp.

Centers for Disease Control and Prevention. *Some common misconceptions about vaccination and how to respond to them*, www.cdc.gov/vaccines/vac-gen/6mishome.htm.

Clark, P.F. 1959. Theobald Smith, Student of Disease (1859–1934). *Journal of the History of Medicine* (October) 490–514.

Dunlop, D.R. 1928. The life and work of Louis Pasteur. *The Canadian Medical Association Journal* 18(3) (March): 297–303.

Fleming, A. 1947. Louis Pasteur. *British Medical Journal* (April 19): 517–522.

Hammarsten, J.F., W. Tattersall, and J.E. Hammarsten. 1979. Who discovered smallpox vaccination? Edward Jenner or Benjamin Jesty? *Transactions of the American Clinical and Climatological Association* 90:44–55.

Hilleman, M.R. 2000. Vaccines in historic evolution and perspective: A narrative of vaccine discoveries. *Vaccine* 18:1436–1447.

Huygelen, C. 1997. The concept of virus attenuation in the eighteenth and early nineteenth centuries. *Biologicals* 25:339–345.

Jenner, Edward. 1798. *An Inquiry Into the Causes and Effects of the Variolae Vaccinae, Or Cow-Pox*, www.bartleby.com/38/4/1.html.

Kaufmann, S.H.E. 2008. Immunology's foundation: The 100-year anniversary of the Nobel Prize to Paul Ehrlich and Elie Metchnikoff. *Nature Immunology* 9(7) (July): 705–712.

Krasner, R. 1995. Pasteur: High priest of microbiology. The American Society for Microbiology. *ASM News* 61(11):575–578.

Li, Y., D.S. Carroll, S.N. Gardner, et al. 2007. On the origin of smallpox: Correlating variola phylogenics with historical smallpox records. *Proceedings of the National Academy of Sciences of the United States of America* 104(40) (October): 15,787–15,792.

Mullin, D. 2003. Prometheus in Gloucestershire: Edward Jenner, 1749–1823. *The Journal of Allergy and Clinical Immunology* 112(4) (October): 810–814.

Nobelprize.org. *Emil von Behring: The Nobel Prize in Physiology or Medicine, 1901*, http://nobelprize.org/nobel_prizes/medicine/laureates/1901/behring-bio.html.

Nobelprize.org. *Ilya Mechnikov: The Nobel Prize in Physiology or Medicine, 1908*, http://nobelprize.org/nobel_prizes/medicine/laureates/1908/mechnikov-bio.html.

Nobelprize.org. *Paul Ehrlich: The Nobel Prize in Physiology or Medicine, 1908*, http://nobelprize.org/nobel_prizes/medicine/laureates/1908/ehrlich-bio.html.

Offit, P.A. 2007. Thimerosal and vaccines—A cautionary tale. *The New England Journal of Medicine* 357(13) (September): 1278–1279.

Pasteur, M. 1881. An address on vaccination in relation to chicken cholera and splenic fever. *The British Medical Journal* (August) 283–284.

Pead, P.J. 2003. Benjamin Jesty: New light in the dawn of vaccination. *The Lancet* 362 (December): 2104–2109.

Plotkin, S.A. 2005. Vaccines: Past, present, and future. *Nature Medicine Supplement* 11(4) (April): S5–S11.

Plotkin, S.A., W.A. Orenstein, and P.A. Offit, eds. 2004. *Vaccines*, 4th edition. Philadelphia: Saunders.

Schwartz, M. 2001. The life and works of Louis Pasteur. *Journal of Applied Microbiology* 91:597–601.

Stern, A.M. and H. Markel. 2005. The history of vaccines and immunization: Familiar patterns, new challenges. *Health Affairs* 24(3) (May/June): 611–621.

U.S. Food and Drug Administration. 2007. *FDA Approves New Smallpox Vaccine*, September 4, 2007, www.fda.gov/consumer/updates/smallpox090407.html.

Chapter 7

Arias, C.A. and B.E. Murray. 2009. Antibiotic-resistant bugs in the 21st century—A clinical super-challenge. *The New England Journal of Medicine* 360 (January 29): 439–443.

Bassett, E.J., M.S. Keith, G.J. Armelagos, et al. 1980. Tetracycline-labeled human bone from ancient Sudanese Nubia (A.D. 350). *Science* 209(4464) (September 26): 1532–1534.

Bentley, S.D., K.F. Chater, A.M. Cerdeño-Tárraga, et al. 2002. Complete genome sequence of the model actinomycete *Streptomyces coelicolor* A3(2). *Nature* 417 (May 9): 141–147.

Brunel, J. 1951. Antibiosis from Pasteur to Fleming. *Journal of the History of Medicine and Allied Sciences* 6(3) (Summer): 287–301.

Capasso, L. 2007. Infectious diseases and eating habits at Herculaneum (1st Century AD, Southern Italy). *International Journal of Osteoarchaeology* 17:350–357.

Centers for Disease Control and Prevention. *Environmental Management of Staph and MRSA in Community Settings*, www.cdc.gov/ncidod/dhqp/ar_mrsa_Enviro_Manage.html.

Centers for Disease Control and Prevention. *Healthcare-Associated Methicillin Resistant* Staphylococcus aureus *(HA-MRSA)*, www.cdc.gov/ncidod/dhqp/ar_mrsa.html.

Chain, E. and H.W. Florey. 1944. The discovery of the chemotherapeutic properties of penicillin. *British Medical Bulletin* 2(1):5–6.

Chain, E.B. 1946. The chemical structure of the penicillins. *Nobel Lecture* (March 20), http://nobelprize.org/nobel_prizes/medicine/laureates/1945/chain-lecture.html.

Chain, E.B. 1979. Fleming's contribution to the discovery of penicillin. *Trends in Biochemical Sciences (TIBS)* (June): 143–144.

Chater, K.F. 2006. Streptomyces inside-out: A new perspective on the bacteria that provide us with antibiotics. *Philosophical Transactions of the Royal Society* 361:761–768.

Davies, J. 2007. Microbes have the last word. European Molecular Biology Organization. *EMBO Reports* 8(7):616–621.

Diggins, F.W.E. 1999. The true history of the discovery of penicillin, with refutation of the misinformation in the literature. *British Journal of Biomedical Science* 56(2):83–93.

Fleming, A. 1944. The discovery of penicillin. *British Medical Bulletin* 2(1):4–5.

Fleming, A. 1945. Penicillin. *Nobel Lecture* (December 11, 1945), http://nobelprize.org/nobel_prizes/medicine/laureates/1945/fleming-lecture.html.

Fleming, A. 1946. *Penicillin: Its Practical Application*. Philadelphia: The Blakiston Company.

Fraser-Moodie, W. 1971. Struggle against infection. *Proceedings of the Royal Society of Medicine* 64 (January): 87–94.

Fridkin, S.K., J.C. Hagerman, M. Morrison, et al. 2005. Methicillin-resistant Staphylococcus aureus disease in three communities. *The New England Journal of Medicine* 352 (April 7): 1436–1444.

Grossman, C.M. 2008. The first use of penicillin in the United States. *Annals of Internal Medicine* 149:135–136.

Hare, R. 1982. New light on the history of penicillin. *Medical History* 26:1–24.

Henderson, J.W. 1997. The yellow brick road to penicillin: A story of serendipity. *Mayo Clinic Proceedings* 72:683–687.

Hobby, G.L. 1951. Microbiology in relation to antibiotics. *Journal of the History of Medicine and Allied Sciences* VI (Summer): 369–387.

Hopwood, D.A. 1999. Forty years of genetics with Streptomyces: From in vivo through in vitro to in silico. *Microbiology* 145:2183–2202.

Kingston, W. 2000. Antibiotics, invention, and innovation. *Research Policy* 29(6) (June): 679–710.

Moellering, R.C., Jr. 1995. Past, present, and future of antimicrobial agents. *The American Journal of Medicine* 99(suppl 6A) (December 29): 11S–18S.

Murray, J.F. 2004. A century of tuberculosis. *American Journal of Respiratory and Critical Care Medicine* 169:1181–1186.

Mycek, M.J., R.A. Harvey, P.C. Champe, et al., eds. 2000. *Lippincott's Illustrated Reviews: Pharmacology*, 2nd ed. Philadelphia: Lippincott Williams & Wilkins.

Naseri, I., R.C. Jerris, and S.E. Sobol. 2009. Nationwide trends in pediatric Staphylococcus aureus head and neck infections. *Archives of Otolaryngology—Head & Neck Surgery* 135(1) (January): 14–16.

Nobelprize.org. *Gerhard Domagk: The Nobel Prize in Physiology or Medicine, 1939*, http://nobelprize.org/nobel_prizes/medicine/laureates/1939/domagk-bio.html.

Nobelprize.org. *Selman A. Waksman: The Nobel Prize in Physiology or Medicine, 1952*, http://nobelprize.org/nobel_prizes/medicine/laureates/1952/waksman-bio.html.

Nobelprize.org. *Sir Albert Fleming, Ernst Boris Chain, and Sir Howard Walter Florey: The Nobel Prize in Physiology or Medicine, 1945*, http://nobelprize.org/nobel_prizes/medicine/laureates/1945/index.html.

Noble, W.C. 1986. The sulphonamides: An early British perspective. *Journal of Antimicrobial Chemotherapy* 17:690–693.

Otten, H. 1986. Domagk and the development of the sulphonamides. *Journal of Antimicrobial Chemotherapy* 17:689–690.

Peláez, F. 2006. The historical delivery of antibiotics from microbial natural products—Can history repeat? *Biochemical Pharmacology* 71(7) (March 30): 981–90.

Penicillin in action. 1941. *The Lancet* (August): 191–192.

Penicillin in America. 1943. *The Lancet* (July): 106.

Saxon, W. 1999. Anne Miller, 90, first patient who was saved by penicillin. *The New York Times* (June 9).

Wainwright, M. and H.T. Swan. 1986. C.G. Paine and the earliest surviving clinical records of penicillin therapy. *Medical History* 30:42-56.

Wainwright, M. 1987. The history of the therapeutic use of crude penicillin. *Medical History* 31:41–50.

Wainwright, M. 2006. Moulds in ancient and more recent medicine (January 20), www.fungi4schools.org/Reprints/Mycologist_articles/Post-16/Medical/V03pp021-023folk_medicine.pdf.

Waksman, S.A. 1952. Streptomycin: Background, isolation, properties, and utilization. *Nobel Lecture* (December 12). http://nobelprize.org/nobel_prizes/medicine/laureates/1952/waksman-lecture.html.

Waksman, S. and D.M. Schullian, eds. 1973. History of the word "antibiotic." *Journal of the History of Medicine and Allied Sciences* (July): 284–286.

Watve, M.G., R. Tickoo, M.M. Jog, and B.D. Bhole. 2001. How many antibiotics are produced by the genus Streptomyces? *Archives of Microbiology* 176:386–390.

Wennergren, G. 2007. One sometimes finds what one is not looking for (Sir Alexander Fleming): The most important medical discovery of the 20th century. *Acta Paediatrica* 96:141–144.

Wiedemann, H.R. 1990. Gerhard Domagk. *European Journal of Pediatric* 149:379.

Williams, D.E. *Patsy's Cure*. American Thoracic Society website, www.thoracic.org/sections/about-ats/centennial/vignettes/articles/vignette4.html.

Chapter 8

Avery, O.T., C.M. MacLeod, and M. McCarty. 1944. Studies on the chemical nature of the substance inducing transformation of pneumococcal types. *The Journal of Experimental Medicine* 79(2) (February 1): 137–158.

Brush, S.G. 1978. Nettie M. Stevens and the discovery of sex determination by chromosomes. *Isis* 69(247) (June): 163–172.

Crow, E.W. and J.F. Crow. 2002. 100 years ago: Walter Sutton and the chromosome theory of heredity. *Genetics* 160 (January): 1–4.

Dahm, R. 2005. Friedrich Miescher and the discovery of DNA. *Developmental Biology* 278(2):274–288.

Dahm, R. 2008. Discovering DNA: Friedrich Miescher and the early years of nucleic acid research. *Human Genetics* 122:565–581.

Dunn, L.C. 1965. Mendel, his work and his place in history. *Proceedings of the American Philosophical Society* 109(4) (August): 189–198.

Fairbanks, D.J. 2001. A century of genetics. *USDA Forest Service Proceedings RMRS-P-21* 42–46.

Feinberg, A.P. 2008. Epigenetics at the epicenter of modern medicine. *Journal of the American Medical Association* 299(11) (March 19): 1345–1350.

Fuller, W. 2003. Who said "helix?" Right and wrong in the story of how the structure of DNA was discovered. *Nature* 424 (August 21): 876–878.

Galton, D.J. 2008. Archibald E. Garrod (1857–1936). *Journal of Inherited Metabolic Disease* 31:561–566.

Glass, B. 1965. A century of biochemical genetics. *Proceedings of the American Philosophical Society* 109(4) (August): 227–236.

Glass, B. 1974. The long neglect of genetic discoveries and the criterion of prematurity. *Journal of the History of Biology* 7(1) (Spring): 101–110.

Goldstein, D.B. 2009. Common genetic variation and human traits. *The New England Journal of Medicine* 360(17) (April 23): 1696–1698.

Harper, P.S. 2005. William Bateson, human genetics and medicine. *Human Genetics* 118:141–151.

Harper, P.S. 2006. The discovery of the human chromosome number in Lund, 1955–1956. *Human Genetics* 119:226–232.

Hartl, D.L. and V. Orel. 1992. What did Gregor Mendel think he discovered? *Genetics* 131 (June): 245–253.

James, J. 1970. Miescher's discoveries of 1869: A centenary of nuclear chemistry. *The Journal of Histochemistry and Cytochemistry* 18(3) (March): 217–219.

Judson, H.F. 2003. The greatest surprise for everyone—notes on the 50th anniversary of the double helix. *The New England Journal of Medicine* 348(17) (April 24): 1712–1714.

Klug, A. 1968. Rosalind Franklin and the discovery of the structure of DNA. *Nature* 219 (August 24): 808–810;843–844.

Klug, A. 2004. The discovery of the DNA double helix. *Journal of Molecular Biology* 335:3–26.

Kohn, D.B. and F. Candotti. 2009. Gene therapy fulfilling its promise. *The New England Journal of Medicine* 360(5) (January 29): 518–521.

Kraft, P. and D.J. Hunter. 2009. Genetic risk prediction—are we there yet? *The New England Journal of Medicine* 360(17) (April 23): 1701–1703.

Lagnado, J. 2005. Past times: From pabulum to prions (via DNA): a tale of two Griffiths. *The Biochemist* (August) 33–35.

Lederman, M. 1989. Research note: Genes on chromosomes: The conversion of Thomas Hunt Morgan. *Journal of the History of Biology* 22(1) (Spring): 163–176.

Macgregor, R.B. and G.M.K. Poon. 2003. The DNA double helix fifty years on. *Computational Biology and Chemistry* 27: 461–467.

Mazzarello, P. 1999. A unifying concept: the history of cell theory. *Nature Cell Biology* (May): 1:E13–E15.

Mendel, G. 1865. Experiments in plant hybridization. Read at the February 8 and March 8, 1865 meetings of the Brünn Natural History Society. www.esp.org/foundations/genetics/classical/gm-65.pdf.

National Institute of General Medical Sciences. *The New Genetics*, http://publications.nigms.nih.gov/thenewgenetics/chapter1.html.

Nobelprize.org. *Thomas Hunt Morgan: The Nobel Prize in Physiology or Medicine, 1933*, http://nobelprize.org/nobel_prizes/medicine/laureates/1933.

O'Connor, C. 2008. Isolating hereditary material: Frederick Griffith, Oswald Avery, Alfred Hershey, and Martha Chase. *Nature Education* 1(1).

Paweletz, N. 2001. Walther Flemming: Pioneer of mitosis research. *Nature Reviews Molecular Cell Biology* (2) (January): 72–75.

Rosenberg, L.E. 2008. Legacies of Garrod's brilliance: One hundred years—and counting. *Journal of Inherited Metabolic Disease* 31:574–579.

Sandler, I. 2000. Development: Mendel's legacy to genetics. *Genetics* 154 (January): 7–11.

Schultz, M. 2008. Rudolf Virchow. *Emerging Infectious Diseases* 14(9) (September): 1480–1481.

Smith, J.E.H., ed. 2006. Introduction. *The Problem of Animal Generation in Early Modern Philosophy*. New York: Cambridge University Press.

Stevenson, I. 1992. A new look at maternal impressions: an analysis of 50 published cases and reports of two recent examples. *Journal of Scientific Exploration* 6(4):353–373.

Sturtevant, A.H. 2001. *A History of Genetics*. Cold Spring Harbor, NY: Cold Spring Harbor Laboratory Press.

Trask, B.J. 2002. Human cytogenetics: 46 chromosomes, 46 years and counting. *Nature Reviews Genetics* 3 (October): 769–778.

Tschermak-Seysenegg, E. 1951. The rediscovery of Gregor Mendel's work: an historic retrospect. *The Journal of Heredity* 42(4):163–171.

U.S. Department of Energy. DOE Joint Genome Institute website. (Numerous articles and background information about research into the human genome), www.jgi.doe.gov.

U.S. Department of Energy. Human Genome Project Information website. (Numerous articles and background information about the Human Genome Project), http://genomics.energy.gov.

U.S. Library of Medicine. The Marshall W. Nirenberg Papers, http://profiles.nlm.nih.gov.

U.S. National Institutes of Health. 2006. Genes or environment? Epigenetics sheds light on debate. *NIH News in Health* (February).

Waller, J. 2003. Parents and children: ideas of heredity in the 19th century. *Endeavor* 27(2) (June): 51–56.

Weiling, F. 1991. Historical study: Johann Gregor Mendel: 1822–1884. *American Journal of Medical Genetics* 40:1–25.

Wiesel, T. Introduction to The Rockefeller University 50th anniversary celebration and reprint of the landmark paper by Oswald Avery, Colin MacLeod, and Maclyn McCarty. www.weizmann.ac.il/complex/tlusty/ courses/landmark/AMM1944.pdf.

Winkelmann, A. 2007. Wilhelm von Waldeyer-Hartz (1836–1921): an anatomist who left his mark. *Clinical Anatomy* 20:231–234.

Chapter 9

Adityanjee, A., Y.A. Aderibigbe, D. Theodoridis, and W.V.R. Vieweg. 1999. Dementia praecox to schizophrenia: the first 100 years. *Psychiatry and Clinical Neurosciences* 53:437–448.

American Psychiatric Association. 2000. *Diagnostic and Statistical Manual of Mental Disorders: DMS-IV-TR,* 4th ed. Text Revision. Washington, DC: American Psychiatric Association.

Baker, A. 2008. Queens man is arrested in killing of therapist. *The New York Times* (February 17).

Baker, A. 2008. Vicious killing where troubled seek a listener. *The New York Times* (February 14).

Balon, R. 2008. The dawn of anxiolytics: Frank M. Berger, 1913–2008. *American Journal of Psychiatry* 165(12) (December):1531.

Ban, T.A. 2007. Fifty years chlorpromazine. *Neuropsychiatric Disease and Treatment* 3(4):495–500.

Berrios, G. 1999. Anxiety disorders: a conceptual history. *Journal of Affective Disorders* 56(2–3):83–94.

Buckley, C. and A. Baker. 2008. Before murder, troubled quest to find mother. *The New York Times* (February 18).

Colp, R., Jr. 2000. History of Psychiatry. In: *Kaplan & Sadock's Comprehensive Textbook of Psychiatry* Vol. II, 7th ed. B.J. Sadock and V.A. Sadock, eds. Philadelphia: Lippincott Williams & Wilkins.

Davies, B. 1983. The first patient to receive lithium. *Australian and New Zealand Journal of Psychiatry* 17:366–368.

Eligon, J. 2008. Bellevue is allowed to medicate suspect. *The New York Times* (June 11).

Estes, J.W. 1995. The road to tranquility: The search for selective anti-anxiety agents. *Synapse* 21:10–20.

Getz, M.J. 2009. The ice pick of oblivion: Moniz, Freeman, and the development of psychosurgery. *TRAMES* 13(63/58),2:129–152.

Goodwin, F.K. and S.N. Ghaemi. 1999. The impact of the discovery of lithium on psychiatric thought and practice in the USA and Europe. *Australian and New Zealand Journal of Psychiatry* 33:S54–S64.

Healy, D. 2001. The Antidepressant Drama. In: *Treatment of Depression: Bridging the 21st Century*. M.M. Weissman, ed. Washington, DC: American Psychiatric Press, Inc.

Konigsberg, E. and A. Farmer. 2008. Father tells of slaying suspect's long ordeal. *The New York Times* (February 20).

López-Muñoz, F., C. Alamo, G. Rubio, and E. Cuenca. 2004. Half a century since the clinical introduction of chlorpromazine and the birth of modern psychopharmacology. *Progress in Neuro-Psychopharmacology & Biological Psychiatry* 28:205–208.

López-Muñoz, F., C. Alamo, G. Rubio, E. Cuenca, et al. 2005. History of the discovery and clinical introduction of chlorpromazine. *Annals of Clinical Psychiatry* 17(3):113–135.

López-Muñoz, F. and C. Alamo. 2009. Monoaminergic neurotransmission: the history of the discovery of antidepressants from 1950s until today. *Current Pharmaceutical Design* 15:1563–1586.

McCrae, N. 2006. A violent thunderstorm: Cardiazol treatment in British mental hospitals. *History of Psychiatry* 17 (March): 67–90.

Middleton, W., M.J. Dorahy, and A. Moskowitz. 2008. Historical conceptions of dissociation and psychosis: Nineteenth and early twentieth century perspectives on severe psychopathology. In: *Psychosis, Trauma and Dissociation*. A. Moskowitz, I. Schäfer, and M.J. Dorahy, eds. West Sussex: John Wiley & Sons, Ltd.

Mitchell, P.B. and D. Hadzi-Pavlovic. 1999. John Cade and the discovery of lithium treatment for manic depressive illness. *The Medical Journal of Australia* 171(5) (September 6): 262–264.

Mondimore, F.M. 2005. Kraepelin and manic-depressive insanity: an historical perspective. *International Review of Psychiatry* 17(1) (February): 49–52.

Nasser, M. 1987. Psychiatry in ancient Egypt. *Bulletin of the Royal College of Psychiatrists* 11 (December): 420–422.

National Alliance on Mental Illness (NAMI). 2007. *Mental Illness: Facts and Numbers* (October), www.nami.org.

National Alliance on Mental Illness (NAMI). 2009. *What is Mental Illness: Mental Illness Facts and Numbers* (October), www.nami.org.

National Institute of Mental Health (NIMH). 2008. *The Numbers Count: Mental Disorders in America*, www.nimh.nih.gov.

Nicol, W.D. 1948. Robert Burton's anatomy of melancholy. *Postgraduate Medical Journal* 24:199–206.

Nobelprize.org. *Egas Moniz: The Nobel Prize in Physiology or Medicine, 1949*, http://nobelprize.org/nobel_prizes/medicine/laureates/1949/moniz-bio.html.

Nobelprize.org. *Julius Wagner-Jauregg: The Nobel Prize in Physiology or Medicine, 1927*, http://nobelprize.org/nobel_prizes/medicine/laureates/1927/index.html.

Panksepp, J. 2004. Biological psychiatry sketched—past, present, and future. In: *Textbook of Biological Psychiatry*. J, Panksepp, ed. Hoboken: Wiley-Liss, Inc.

Ramchandani, D., F. López-Muñoz, and C. Alamo. 2006. Meprobamate—tranquilizer or anxiolytic? A historical perspective. *Psychiatric Quarterly* 77(1) (Spring): 43–53.

Roffe, D. and C. Roffe. 1995. Madness and care in the community: a medieval perspective. *British Medical Journal* 311 (December 23): 1708–1712.

Sadock, B.J. and V.A. Sadock. 2007. *Kaplan & Sadock's Synopsis of Psychiatry: Behavioral Sciences/Clinical Psychiatry*, 10th ed. Philadelphia: Lippincott Williams & Wilkins.

Safavi-Abbasi, S., L.B.C. Brasiliense, R.K. Workman, et al. 2007. The fate of medical knowledge and the neurosciences during the time of Genghis Khan and the Mongolian Empire. *Neurosurgical Focus* 23(1) (July): 1–6.

Scull, A. 1983. The domestication of madness. *Medical History* 27:233–248.

Shen, W.W. 1999. A history of antipsychotic drug development. *Comprehensive Psychiatry* 40(6) (November/December): 407–414.

Tone, A. 2005. Listening to the past: history, psychiatry, and anxiety. *Canadian Journal of Psychiatry* 50(7) (June): 373–380.

Weiner, D.B. 1992. Philippe Pinel's "Memoir on Madness" of December 11, 1794: a fundamental text of modern psychiatry. *American Journal of Psychiatry* 149 (June): 725–732.

Whitrow, M. 1990. Wagner-Jauregg and fever therapy. *Medical History* 34:294–310.

World Health Organization (WHO). 2001. *Mental Health: New Understanding, New Hope*, www.who.int/whr/2001/en/whr01_en.pdf.

World Health Organization (WHO). 2006. *Dollars, DALYs and Decisions: Economic Aspects of the Mental Health System*, www.who.int/mental_health/evidence/dollars_dalys_and_decisions.pdf.

World Health Organization (WHO) and World Organization of Family Doctors (Wonca). 2008. *Integrating mental health into primary care: a global perspective*, www.who.int/mental_health/policy/Mental%20health%20+%20primary%20care-%20final%20low-res%20140908.pdf.

Chapter 10

Barnes, P.M., B. Bloom, and R. Nahin. 2008. *CDC National Health Statistics Report #12. Complementary and Alternative Medicine Use Among Adults and Children: United States, 2007*. National Health Statistics Reports (National Center for Health Statistics, Centers for Disease Control and Prevention) 12 (December 10): 1–24.

Benedic, A.L., L. Mancini, and M.A. Grodin. 2009. Struggling to meditate: contextualising integrated treatment of traumatized Tibetan refugee monks. *Mental Health, Religion & Culture* 12(5) (July): 485–499.

Bivens, R. 2007. *Alternative Medicine? A History*. New York: Oxford University Press, Inc.

Chopra, A. and V.V. Doiphode. 2002. Ayurvedic medicine: core concept, therapeutic principles, and current relevance. *Medical Clinics of North America* 86(1) (January): 75–89.

Daniloff, C. 2009. Treating Tibet's traumatized. *Bostonia: The Alumni Magazine of Boston University* No. 3 (Fall): 13.

DeVocht, J.W. 2006. History and overview of theories and methods of chiropractic. *Clinical Orthopaedics and Related Research* 444 (March): 243–249.

Dunn, P.M. 2003. Galen (AD 129–200) of Pergamun: anatomist and experimental physiologist. *Archives of Disease in Childhood (Fetal and Neonatal Edition)* 88(5) (September): F441–F443.

Eisenberg, D.M., R.C. Kessler, C. Foster, et al. 1993. Unconventional medicine in the United States: prevalence, costs, and patterns of use. *The New England Journal of Medicine* 328(4) (January 28): 246–252.

Eisenberg, D.M., R.B. Davis, S.L. Ettner, et al. 1998. Trends in alternative medicine use in the United States, 1990–1997. *Journal of the American Medical Association (JAMA)* 280(18) (November 11) 1569–1575.

Homola, S. 2006. Chiropractic: history and overview of theories and methods. *Clinical Orthopaedics and Related Research* 444 (March): 236–242.

Jonas, W.B., T.J. Kaptchuk, and K. Linde. 2003. A critical overview of homeopathy. *Annals of Internal Medicine* 138(5) (March 4): 393–399.

Kaptchuk, T.J. and D.M. Eisenberg. 2001. Varieties of healing. 1: Medical pluralism in the United States. *Annals of Internal Medicine* 135(3) (August 7): 189–195.

Kaptchuk, T.J. and D.M. Eisenberg. 2001. Varieties of healing. 2: A taxonomy of unconventional healing practices. *Annals of Internal Medicine* 135(3) (August 7): 196–204.

Keating, J.C., Jr., C.S. Cleveland, III, and M. Menke. 2004. *Chiropractic History: A Primer.* Davenport, Iowa: Association for the History of Chiropractic.

Khuda-Bukhsh, A.R. 2003.Towards understanding molecular mechanisms of action of homeopathic drugs: an overview. *Molecular and Cellular Biochemistry* 253:339–345.

Kilgour, F.G. 1961. William Harvey and his contributions. *Circulation* 23 (February): 286–96.

Knoll, A.M. 2004. The reawakening of complementary and alternative medicine at the turn of the twenty-first century: filling the void in conventional biomedicine. *Journal of Contemporary Health Law and Policy* 20(2) (Spring): 329–66.

Leis, A.M., L.C. Weeks, and M.J. Verhoef. 2008. Principles to guide integrative oncology and the development of an evidence base. *Current Oncology* 15(Suppl 2): S83–S87.

Lüderitz, B. 2009. The discovery of the stethoscope by T.R.H. Laënnec (1781–1826). *Journal of Interventional Cardiac Electrophysiology* (July 29) www.springerlink.com/content/c10r07519l4370h7/.

Marketsos, S.G. and P.K. Skiadas. 1999. The modern Hippocratic tradition: some messages for contemporary medicine. *Spine* 24(11):1159–1163.

Meeker, W.C. and S. Haldeman. 2002. Chiropractic: a profession at the crossroads of mainstream and alternative medicine. *Annals of Internal Medicine* 136(3) (February 5): 216–227.

Micozzi, M.S. 1998. Historical aspects of complementary medicine. *Clinics in Dermatology* 16:651–658.

Nahin, R.L., P.M. Barnes, B.J. Stussman, and B. Bloom. 2009. Costs of complementary and alternative medicine and frequency of visits to CAM practitioners: United States, 2007. *National Health Statistics Reports* (National Center for Health Statistics, Centers for Disease Control and Prevention) 18 (July 30): 1–14.

National Center for Complementary and Alternative Medicine (NCCAM). (National Institutes of Health, U.S. Dept. of Health and Human Services.) Wide variety of background information on CAM therapies, research, grants, training, and news. http://nccam.nih.gov.

Nestler, G. 2002. Traditional Chinese medicine. *Medical Clinics of North America* 86(1) (January): 63–73.

Nutton, V. 2002. Logic, learning, and experimental medicine. [Galen] *Science* 295 (February 1): 800–801.

Nutton, V. 2004. *Ancient Medicine*. London: Routledge.

O'Malley, C.D. 1964. Andreas Vesalius 1514–1564: In Memoriam. *Medical History* 8 (October): 299–308.

Rakel, D. and A. Weil. 2007. Philosophy of Integrative Medicine. Chapter 1. In: Rakel. *Integrative Medicine*, 2nd ed. Philadelphia: Saunders Elsevier.

Rees, L. and A. Weil. 2001. Integrated medicine imbues orthodox medicine with the values of complementary medicine. *British Medical Journal* 322 (January 20): 119–120.

Sakula, A. 1981. RTH Laënnec 1781–1826. His life and work: a bicentenary appreciation. *Thorax* 36:81–90.

Scherer, J.R. 2007. Before cardiac MRI: René Laennec (1781–1826) and the invention of the stethoscope. *Cardiology Journal* 14(5):518–519.

Schultz, S.G. 2002. William Harvey and the circulation of the blood: the birth of a scientific revolution and modern physiology. *News in Physiological Sciences* 17 (October): 175–180.

Subbarayappa, B.V. 2001. The roots of ancient medicine: an historical outline. *Journal of Biosciences* 26(2) (June): 135–144.

Tan, S.Y. and M.E. Yeow. 2003. Ambroise Paré (1510–1590): the gentle surgeon. *Singapore Medical Journal* 44(3):112–113.

Tan, S.Y. and M.E. Yeow. 2003. Andreas Vesalius (1514–1564): father of modern anatomy. *Singapore Medical Journal* 44(5):229–230.

Tan, S.Y. and M.E. Yeow. 2003. William Harvey (1578–1657): discoverer of circulation. *Singapore Medical Journal* 44(9):445–446.

Thomas, L. 1995. *The Youngest Science: Notes of a Medicine-Watcher*. London: Penguin Books Ltd.

Todman, D. 2007. Galen (129–199). *Journal of Neurology* 254:975–976.

Vickers, A.J. and C. Zollman. 1999. Homeopathy. *BMJ* (clinical research ed.) 319 (October 23): 1115–1118.

Willms, L. and N. St. Pierre-Hansen. 2008. Blending in: Is integrative medicine the future of family medicine? *Canadian Family Physician* 54 (August): 1085–1087.

Index

A

B

C

D

E

F

G

H

I

J-K

L

U-V

W

X-Y-Z

In an increasingly competitive world, it is quality of thinking that gives an edge—an idea that opens new doors, a technique that solves a problem, or an insight that simply helps make sense of it all.

We work with leading authors in the various arenas of business and finance to bring cutting-edge thinking and best-learning practices to a global market.

It is our goal to create world-class print publications and electronic products that give readers knowledge and understanding that can then be applied, whether studying or at work.

To find out more about our business products, you can visit us at www.ftpress.com.